ANIMAL SOCIETIES

Remy Chauvin

Homo Sapiens has existed for 150,000 years,
ants and bees for 40 million at least, and
termites for more than 200 million. Small
wonder that so many of the social habits and
techniques of the animal kingdom—
particularly among the insects—remain a
complete mystery to man. Slavery among the
ants, the 'language' of bees, the migration of
birds and the suicide of the lemmings.

Professor Chauvin explores many other
features of animal societies including the
demarcation of territory, the 'pecking order',
the nature of communication, ceremonial
courtship and social attitudes to the very
young or very old—and makes a number of
astonishing comparisons with human society.

ANIMAL SOCIETIES is an ideal companion
to the work of such writers as Konrad Lorenz,
Robert Ardrey and Desmond Morris.

"Admirable"—*Guardian*
"Fascinating"—*Sun*
"Compelling"—*Sunday Times*
"Important"—*Sunday Telegraph*
"Profound"—*Daily Telegraph*
"Excellent"—*Times Educational Supplement*

ANIMAL SOCIETIES

Animal Societies

REMY CHAUVIN

SPHERE BOOKS LIMITED
30/32 Gray's Inn Road, London, WC1X 8JL

First published in Great Britain in 1968 by
Victor Gollancz Ltd.
© Librairie Plon, Paris, 1963
English translation © Victor Gollancz Ltd. 1968
First Sphere Books edition 1971

TRADE
MARK

Set in Linotype Times

Printed in Great Britain by
Hazell Watson & Viney Ltd
Aylesbury, Bucks

CONTENTS

INTRODUCTION

By starting this book with an account of insect societies the reader will soon appreciate how remote these wonderful, but mechanical, creatures are from man. "Inhuman" is not the word to use; "non-human" or "extra-human" might be better. In fact they are hardly societies, but true organisms, as I shall explain later on: the hive or the ants' nest is the real unit, the isolated bee or ant is nothing more than an abstraction ... a less paradoxical idea than might at first appear. There are considerable arguments for it, particularly the utter dependence of the individual on the group: separated from its own kind a bee or ant can do no better than die within a few days or hours. We are indeed dealing with another world, one so far away from us that it might have fallen from another planet.

But if we leave this world and turn to that of the vertebrates we feel "at home"; we are back in a world we know. The conjugal disputes and dances of birds, the social organization of a group of macaque monkeys may even disturb us by their closeness to human behaviour. No one denies nowadays that our behaviour is deeply rooted in animal origins—but when we see just how deeply, even in the ways of life we think most typically ours. . . . It makes one ponder. However, it is our world; it is not the too well oiled machine of the insects. How many hours have I spent looking at ants' nests in the woods; a spectacle both fascinating and, from certain aspects, frightening! But to hear a nightingale, or see the courting swans, brings me back to our own gentle planet, mother of the human race.

Nevertheless I am an entomologist, and I feel a certain disappointment when dealing with vertebrates. They seem too simple and their habits too gross. What are they, then, these primates who build no houses, raise no cattle, grow no fungi, do not collect honey for storage? Ants and bees have been doing this for millions of years.

The disappointment stems from the fact that with vertebrates, even primates, we are in a world just before the Stone Age, whilst in the case of the social insects we are studying a *civilization* infinitely older than that of man. It must be understood that we are talking of an *insect civilization*, still utterly different in kind from our own. But if by "civilization" we mean the development of complicated social organization, the carrying out of works in common, the methodical care of the young by the community, and the division of labour among specialized groups, then undoubtedly we have the right to use this word. Obviously man does all this and more, by means of fundamentally different methods. This difference of methods is, moreover, the strangest aspect of the study of entomological sociology. The absorption of the individual in society is here pushed to its extremes; it has scarcely started with us. This is perhaps why adventurous biologists, such as Escherich, put forward termite and ant societies as models for man: the internal logic so cruelly demonstrated there fascinates them.

But it is only an *insect logic* with which we are dealing. In reality—and this is not a hypothesis, it is an established reality —the process of evolution in all the phyla is towards a higher development of the nervous system, a greater psychism. Oysters are not very lively, but nevertheless the molluscs culminate in the Cephalopodae, such as the octopus, with a large brain, complex instincts and agile tentacles that more than make up for the absence of hands. With birds, as we shall see, social instinct has developed and reaches high in the scale of psychism; unfortunately we know almost nothing of truly social birds. With mammals, a small group poor in both species and numbers, it is man, alone among animals, who has broken through the barrier of consciousness.

But terrestrial life has not invested most heavily in man, but in *insects*. A million and a half species are known, and without doubt there are three times as many more to be discovered. Several thousand new species are described each year. There are many more species of flies in France than there are of mammals in the whole world, and these flies differ far more among themselves than do mouse and elephant. In short, at least 80 per cent of animal species are insects. They have not escaped the common law, the ascent towards greater psychism. But with insects there is a considerable obstacle here, their small size, which imposes a fatal limit on the number of nerve cells their brains can hold. How can this be overcome? Insect societies have found the answer, by interconnecting all these small individual brains in ways we are just beginning to understand. This allows an amazing development to take place—agriculture, cattle-raising, harvesting and food storage, and in addition war and slavery.

But after that, progress stops. Why? Why did man dominate and not the ant? The next step the insects had to take seems sometimes so small. But they have just stayed where they were. No doubt science will tell us why one day. Things may have gone differently on other planets, who knows?

In any case this puts into perspective any premature judgements between man and ant. In truth common sense is against such comparisons because they fail to take notice of the enormous and very ancient divergence which separates mammals and insects. Ants and bees were already in existence 40 million years ago at least, and scarcely differ from those we know today. As for termites, everything leads to the supposition that they are more than 200 million years old. And *Homo sapiens* has had hardly 150,000 years of existence . . . a complete difference of kind separates us. It is this that I try to show in the following pages, by comparing the absolutely non-human societies of the insects, outside our world, with societies that are merely "infrahuman", or "parahuman" (in any case not "anti-human"): those of the birds and primates.

They belong to our world, and we see there, not without emotion, the blurred and indistinct outlines of human behaviour.

PART ONE

INSECT SOCIETIES

CHAPTER ONE

THE BEE

THE HISTORY OF THE BEE: SOME REASSURING ASPECTS

THE BEE is an old friend of man. Man has had bees near him from the depths of time, and no doubt because of that thinks he knows them. The mass of superstitions and folklore accumulated around bees defies description. It seems to be the creature about which most has been written. With the Egyptians the soul leaving the body often took the shape of a bee. But who can say why Artemis of Ephesus is sometimes shown as a bee? And why is it that throughout the West there is a firm belief that the bees must be told of a death in the family? Moreover, they must be addressed politely, with flattering words; otherwise they may be upset and abandon the hive. With the Franks the bee—which each soldier had to cultivate— became a kind of national emblem, later adopted by Napoleon. Still today there is an atmosphere of mystery surrounding bee-keeping. The beekeeper is regarded with a respectful fear, of which he is not a little proud. He learnt the job himself, or possibly an old man told him his secrets. The atmosphere of mystery in which he acquired his art makes him impervious to the progress of science. Everyone knows that something learned from one's grandfather, or an old uncle, over a glass of mead is far more valuable than a lot of stuff from those chaps in white overalls without enough wax on their fingers to make them worth listening to. This explains why progress is so slow. Until very recent times our methods with hives were the same as those of the neolithic age. The big inventions (centrifugal honey extractor and, above all, movable frame hives) are barely a century old. And it is even more recently that we have tried selecting strains of bees as we do with other domestic animals, using artificial insemination as for cows.

The beekeeping profession is as mysterious as its traditions.

It refuses obstinately to supply or consider statistics or to listen to an apicultural instructor; it is only by crosschecking that one can obtain any figures—and these are somewhat startling. When an area is honey-bearing a practised eye will always discover hives in the coppices and folds of the countryside; or, by following the flight of the worker bees, guess at the presence of fifty colonies behind some wall. It is probable that in France there is an average of at least 20,000 hives in each department; some put it at double this figure. Honey is collected by tens of thousands of tons (at least 25,000 tons counting only the professional beekeepers). I used to know a beekeeper who himself alone had 3,000 hives and got, good year, bad year, 30 to 40 tons of honey, thanks to what was almost a factory system. A great many of them own a thousand colonies. It may be noted in passing that a large part of the honey in Alsace and Germany is honey-dew—that is, it is basically the excretion of the fir-tree aphids: these aphids are a matter of great interest to ants on the one hand (a large ants' nest will need around a hundred kilos of aphid honey-dew a year, which quantity will scarcely suffice for its millions of workers) and bees on the other. The Germans are particularly fond of this "fir-honey" and the crop reaches some 25,000 to 30,000 tons, according to the season.

These extraordinary figures become more readily explicable if one stops thinking of the bee in isolation, as is too often done, and considers the hive as a whole.

THE "DISTURBING" ASPECT OF BEES

Many biologists, myself among them, are tending more and more to alter their concept of the bee as an isolated insect. What sort of an individual is it that can live only a few hours out of contact with its fellows? Surely some essential factor is missing, as in the case of those tissue cultures which degenerate and take the form of ordinary connective tissue when separated from the parent body. Suppose the bee to be no more than an abstract idea in our minds, suppose insect societies to be not societies but organisms of which the bees, the ants, the termites are the cells? This would merely presuppose that the intercellular relationships are less well defined than in our own bodies: the "cells" could detach themselves temporarily from the organism to go and search for food, defend the colony against attack, etc. . . . And all the comparisons that have been made, or could be made, between human society and that of bees would come from a basic misunderstanding of the true nature of bees. I am not unaware that this is a somewhat startling concept for the layman. It is a theme I will deal with as this book proceeds and you will see that it is on a more solid foundation than might be imagined at first sight.

THE FAMILY OF BEES

There is a fairly large number of species of solitary bees. They can often be seen in spring, collecting pollen from dandelions. They look like a slightly smaller version of the honey bee (when they are probably *Andrena* sp.) or are very small (when they are likely to be *Halictus* sp.). These solitary bees raise their larvae on a mixture of honey and pollen, in tunnels or hollow stalks, after which the mother bee dies. The *Andrena*, for instance, come under this heading. As to *Halictus*, it has recently been found that they are social and make small colonies on the sides of roads where the queen bee, surrounded by some of her daughters, takes care of her offspring at the end of a branching gallery.

The enormous and splendid bumble-bees, buzzing in the spring like a lot of helicopters, are also social. The female founder takes over a rodent's burrow in the spring and starts to lay eggs there, in a waxy bag which she enlarges as the larvae grow. The females which develop will soon be putting nectar and pollen into coarse cells made of a mixture of wax and earth. Large females, born at the end of summer, pass the winter in the wild and start new colonies the following spring. The old queen dies, and the nest is at once invaded by all sorts of creatures drawn there by the attractive smell of honey and wax.

These then are the colonies I would call imperfect, since they cannot maintain themselves the year round. In South America, on the other hand, we find bees with permanent nests, the *Melipona* spp.; their chief peculiarity, in the view of the ordinary man at least, is that they have no sting. It is true that they bite fiercely any intruder who disturbs their nest, which is almost as disagreeable as a sting. There are several species, varying in size from that of a fly to that of our own bees. Their habits are very variable and little known. Differences exist between their nests and those of our own bees: the comb is horizontal instead of being vertical, and the cells open at the upper end (with wasps, who also have horizontal combs, the cell opens at the lower end). Both honey and pollen are stored in big wax balls kept at some distance from the comb. The Maya got honey from these *Melipona*. The hive was protected by a special god and these Indians had developed an apicultural technique not unlike our own, using artificial swarming, etc. It was much later that our own bee was introduced to the American continent, where previously it was quite unknown.

THE NEAR RELATIVES OF THE HONEY-BEE

The bee is not the only creature in the genus *Apis*; a few other species, all Asiatic, keep our bee company. For example, *Apis florea*, a very small species from India, makes its nest in the open air, attaching it to a branch. Unfortunately the honey smells good and the insatiable ants will risk anything to get hold of it. But usually in vain, for the bees cover the branch here and there with bands of sticky resin. Never mind: the ants then perform their favourite manœuvre, bringing bits of twig and debris to form a bridge. At least, that is their intention: but no sooner are the twigs in place than the little bees cover them with a new layer of resin. One assumes that the ants eventually get tired of the game, since *Apis florea* still flourishes.

The giant of the genus is *Apis dorsata*, almost as big as a hornet, which makes a huge comb, at times as big as a door, and also attaches it to a tree branch. This bee is very fierce and very dangerous too on account of its weapon, more like a dagger than a sting. This has not discouraged the Indians, who have often tried to domesticate the creature; but this bee absolutely refuses to be shut in a hive and deserts it at the first opportunity. It is said that certain natives of Indonesia manage to collect *dorsata* honey; the beekeeping costume is much simplified, since they go almost naked; one is lost in speculation as to how they escape death (in any case they get badly stung); perhaps they rub themselves with some repellent, to stop things going too far.

APIS MELLIFICA

Let us first run over a few elementary facts, without which it would be impossible to continue this book. The hive is mono-gynous; a second queen is only tolerated under exceptional conditions, and never permanently. The queen should lay 1,500 to 2,000 eggs every day, which includes the night, for there is no real nocturnal sleep among social insects; the workers merely stop foraging. In our climates egglaying stops towards the month of October and is not taken up again until around 15 February. This recommencement is very slow and only a few eggs per day are laid. The first eggs are fertilized and give rise to worker females: towards the months of May–June males appear, com-ing from unfertilized eggs laid by the queen in rather larger cells; from this it follows that the queen can determine the sex of the eggs at will. The reaction is thought to be due to an abdominal reflex triggered by the size of the cell; according to whether the abdomen is more or less contracted it brings about, or does not bring about, the fertilization of the egg. To do this there is a special apparatus called the sperm pump, situated behind the sperm reservoir where a sufficient supply of sperm is kept to fertilize the eggs for several years. There are hardly more than a few hundred males in a hive, whilst there are 40,000 or 50,000 workers, or barren females. In May–June the bees prepare for swarming. Then a few young larvae get special food—the famous royal jelly—prepared by the pharangeal and mandibular glands of the bees, a food which is in fact given to all bee larvae for the first few hours of life. But the workers scarcely receive 2 or 3 milligrams, and after the first stage they only get a coarse diet of bee bread, about which little is known. The royal larvae, by contrast, rejoice in from 100 to 300 milli-grams of thick broth resembling yoghourt; this is the royal jelly, the sole diet on which they develop. They become female adults with fully functional ovaries. As to the males, we are not so well informed about their diet; it is probable that they receive a mixture of honey and pollen. The young queen, once adult, passes a few days in the hive where the workers do not pay much attention to her until she is fertilized (we may note in passing that it used to be thought that the queens only mated once; we now know that the sperm from a single male would not be enough to last a queen for years and that she really mates with from five to ten males).

Then the colony is seized with an extraordinary excitement, during which the temperature can approach 40°C. The workers push the old queen towards the exit, for it is she who will leave with the swarm: moreover she seems to leave with reluctance. When at last she flies off she takes half the population with her, as Sendler has shown, the halving being almost exact, within a few tens of grams. Whilst the new queen stays in the hive the swarm goes off to install itself in a new site favoured by the scouts.

When winter comes the males are massacred, or simply thrown out where the cold quickly kills them. The queen stops laying and the bees get into a close mass, like a bunch of grapes, where the temperature is kept at 12–15°C even if it is freezing outside. And this temperature control is steadily maintained as long as there are supplies of honey.

ON THE SOCIAL PHYSIOLOGY OF BEES: TEMPERATURE CONTROL

Bees differ from all other insects in their ability to maintain the temperature of the hive at an almost constant figure. It is not the case that insects are necessarily at the same level of temperature as their surroundings, as is often supposed. At periods of muscular activity the temperature of the thorax, where the locomotor muscles are found, generally rises considerably: up to 35° or 40°C, for example, in the big sphinx moths after a fairly long flight. But bees have carried heat production to a level unknown among other variable temperature animals. Nevertheless, as Esch, of Munich, has remarked, even in isolation the bee possesses the faculty of maintaining its body temperature much better than can other insects, as long as it has supplies of sugary material. But the hive, in its role of "super-organism", has faculties far beyond this.

In the brood comb, that central part of the comb where the eggs, the larvae and the nymphs are found, the temperature during the breeding season is constant at 33–34°C, again provided that there are sufficient supplies of sugars; we know little of how they produce this heat. It is curious to note in passing that the temperature of the thorax rises considerably during the dances on the comb, which will be discussed below, even among those bees which just follow the dancer. When the bee "heats", according to Esch, the potential thoracic muscle action reaches a volume and frequency characteristic of flight, though in fact the wings are not moved. When the heat is excessive, bees (and also wasps) collect water and evaporate it on the comb; others form a line with their posteriors towards the exit and by beating their wings give rise to a current of air which quickly removes the unwanted calories. Finally, if the temperature continues to rise, large numbers of bees come out of the hive to remain motionless on the outside of it, near the entrance. These activities are referred to by beekeepers as "fanning".

When the off-season comes the bees who are getting ready to face it are appreciably different from the summer bees; the latter have few reserves of food and a short life; the former have a considerable stock of protein and fat, enough to keep them alive for six months or more. They get together in the middle of the nest in a close, almost immobile cluster, in the centre of which the temperature will scarcely fall below 13°C whatever the outside weather. In the centre of the mass a small hot zone is found, a few centimetres across, which at times experiences considerable and very irregular temperature changes. By fixing very fine thermocouples to the thoraxes of the bees on the outside of the swarm, Esch was able to observe their changes of temperature at his leisure. Bees often stayed on the outside for a long time without apparently suffering any inconvenience from the very low temperature. Then finally they withdrew into the body of the swarm where it was warmer. The sudden rises in temperature produced appear to be related to the intake of syrup; in any case the temperature of the thorax rises quickly from the moment the bee starts to ingest sugar. After this those who have fed distribute the food to the others, as is the custom among all insect societies (see below), and the thoraxes of the recipients then start to "heat up" to the same level as those of the donors; to the extent that thermal control in winter depends on food circulation.

THE SEARCH FOR FOOD

Everbody knows that bees suck nectar from flowers; but, as we have just seen, they do not despise a less noble substance, honey-dew, the excretions of certain aphids. They do not, however, limit themselves to nectar; they are also interested in water (more or less pure), propolis and pollen.

I think that here I must give a few facts about the structure and function of flowers because I have often been surprised how few laymen understand the nature of nectar and pollen. Nectar is a sugary secretion given off by glands usually placed at the base of flower petals. One can easily see them by using a small hand lens on an ordinary colza flower. Between the big yellow petals you can see shiny green swellings; these are the nectaries. I take colza as an example because it is the richest honey-producing plant in the whole region north of the Loire. Bee-keepers put their hives out in it in early spring and, if the weather. is not too bad, a curious sight develops. An absolute curtain of bees flows from the flowers to the hives, which increase in weight by two to three kilos *per day*. One can hardly even inspect the hives, for the mere lifting out of a comb gives rise to a flood of fresh nectar dripping from all the cells. The visits to these hives in early spring are one of the best memories I have of the time when I ran the apicultural laboratory: the yellow sea of colza, the young buds starting up, the sky looking new and fresh; and mixed with the wild birdsong the murmuring of the bees, reminding me of Virgil and the solemn syllables of the mother tongue, "*Tantus amor florum et generandi gloria mellis . . .*"

But how do bees find the nectar? By means of scouts sent out for the purpose, who pass on this precious information to their co-workers by methods we will discuss later on. For the present it should be noted that it is odour and flower colour which guide the bees. Thanks to the remarkable work of von Frisch and his school we have, perhaps, an even better understanding of the functions of sight and smell in bees than in man. And, as you will see, the methods which taught us are admirably simple ones.

But first of all let me introduce you to the inspirer of this research, the greatest experimental biologist perhaps since Pasteur, Professor Karl von Frisch.

It is a thrilling experience to meet these great scientists, something that has happened to me three or four times in my life. And how different they are one from the other! Von Frisch is the German Herr Professor, cold, master of himself, always deep in thought, who, over a period of forty years, has never left the subject of bees. I do not know why, but he reminds me of certain Buddhist *clokas* who praise the "Buddhisatva, the all-wise". By mental ability and an imperturbable concentration, he has reached, as he himself says, right inside his subject. He "feels himself a bee". He knows how a bee will react under any given circumstance. How different is his compatriot, the famous Konrad Lorenz, who has given new impetus to the study of animal psychology; enormous, exuberant, with rumpled hair and beard, talking and joking in three or four languages, always glad to lead you through the maze of the behaviour of wild geese. Or then Grassé, my teacher, such an enthusiast on termites and with such a deep knowledge of the animal world. Such men see the relationships in living nature before other men do; they merely have to confirm by experiment what their intuition has already told them.

In order to find out what bees see and smell all that need be done is to use a very simple technique, that of studying their ability to learn, for worker-bees can be trained very easily. Let us suppose that we show the bees a series of plates, one of them scented with essence of lavender, and that they only get syrup on the lavender plates. Now let us change the arrangement of the plates, but without putting any syrup on the lavender plate this time. If the bees always come down on the lavender plate without fail because it is because they recognize the lavender smell. In a second test we can replace the lavender smell by another rather similar one; the bees then either will not or will concentrate on the new smell. In the first case one can conclude that they mix up the two odours and in the second that they can distinguish them. Thus by steadily working on, varying the perfumes and their concentration, all the olfactory possibilities possessed by bees can be explored. The question of colour is rather more subtle; the question is whether bees see the world in colour, as we do, or in black and white. Let us take a big surface, about a metre square, and cover it with squares in all the different shades of grey, from the lightest to the darkest, all in a random arrangement. Suppose we want to study the colour blue. On a blue square, placed anywhere among the grey squares, we put a little honey to attract a worker. As soon as the bee has finished feeding we move the blue square to some other place at random among the grey squares; there will certainly be a grey square equal in depth of tone to the blue; that is, if one photographed the board in black and white there would be a grey square of the same tone as the blue one. One of two things will now happen; the bee will either find the square again in spite of many permutations, or it will not. In the first case this means that the bee can distinguish blue as a colour, and in the second that the world appears to it as a black and white one. From such experiments we can say that the bee does not see red, but distinguishes ultra-violet; for two white squares can appear different to it according to whether the paint absorbs or reflects the ultra-violet light.

The sense of smell in bees does not differ very much from our own, in spite of the very different anatomy, except that the bee is more sensitive to flower odours, as may be imagined, and it is not repelled by certain smells of decay (as we shall see later). Smell, then, helps the scouts to find a field of flowers as does also sight. But think how the countryside must appear to a bee, which sees no shades of red but does see ultra-violet! Poppies (which interest them greatly because of their pollen) would seem to them to be black, except that they reflect a great deal of ultra-violet. Foliage does not reflect much and must seem to be all shades of light grey, while the flowers stand out from it much more strikingly than they do for human eyes. "White" flowers are not white to them, owing to the ultra-violet component possessed by them all in this climate; everything, down to the wind-blown movement of the flowers, helps the bee to find them, for all moving objects attract them. The worker-bees, alerted by the famous bee dances, of which we shall speak later on, follow a well-marked aerial pathway in order to reach their goal, as has been shown by Lecomte's work (see p. 68); one can mark such a route on a map and often it is the same from one year to the next. It is strange to note how they seem to be fixed as it were on railway lines and that they will neglect, for instance, a fine chestnut in flower on the edge of their route. They go straight to the place pointed out by the scout and towards nothing else.

Pollen, made up of little grains of dust with finely sculptured surfaces, each one only a few thousandths of a millimetre in diameter, is the male element in flowers and is produced by the anthers, the slender stalks which surround the pistil. The pistil is a column situated in the centre of the corolla. For the flower to produce a fruit, the pollen grains must fall on the pistil and must germinate there like a cereal grain, and the long tube which then grows out, with the sexual elements at its extremity, must reach the ovary at the base of the pistil. A startling figure should now be noted. Pollen is rich in nutritive substances, a little like beer yeast, and people have dreamed of making a dietetic food from it; a small device, the pollen trap, put in front of the hives, makes it easy to collect 100 grams of pollen per hive per day during the summer. Thus French beekeepers now produce several tens of *tons* and could produce much more if it were wanted. But let us get back to flowers; the case mentioned above, where the pollen falls directly on to the pistil (self-fertile flower), is not so very common. Very often flowers are wind pollinated (anemophilous flowers) and still more often the flowers need the help of an insect which, flying from flower to flower looking for nectar, has every chance of rubbing its pollen-dusted body against the pistil (entomophilous flowers). This is all the more necessary in the case of the so-called self-sterile flowers, such as the apple, which cannot be fertilized with their own pollen, but must have that of a different variety. Sixty-five per cent of the apples we eat we owe to bees, and the remainder to other insects. Things being as they are, the tree-grower can not do without bee-hives, and nor can the lucerne grower. But here the matter is far more complex. As soon as a bee tries to introduce its proboscis into a lucerne flower the stigma springs out from the petals and strikes the insect, I would say on the chin, if the insect had such a feature. The surprised bee jumps backwards, probably never to return to such an alarming flower. Thus other insects have been sought which will not be upset by such a movement, or which will get at the nectar in another way. For instance, there are the bumble-bees, and efforts are being made at the moment to increase these insects; this is why there is in America a bumble-bee apiculture at the present moment, exclusively for the purposes of pollination. It is developing into a true industry, and huge lorries can be seen carrying hundreds of hives hired by the producers of lucerne seed. As the amount of seed produced rises in direct proportion to the number of worker-bees the fields are methodically over-populated; and because of the competition the bees are not able to collect enough honey to provide for their keep. But supplementary syrup is given them to make up for the deficiency. They are not being asked to produce honey, but seed.

DRINKERS OF DEW?

The bees may be chaste, must be so indeed, because their sexual glands are imperfectly developed; but as to drinking dew, as the poets have liked to depict, they are far from doing that! What the bees prefer is to quench their thirst at urinals and, if there are none, there is no stagnant water too foul for them. Quite true! Bees in fact prefer water containing mineral salts, and, according to some, traces of indol and scatol, foul-smelling products of rotting organic material, may positively attract them. They are guided to pools by their antennae, which can sense humidity. . . . But the number of water carriers is generally very small and one might well ask what regulates their movements. It seems, according to some recent German work, that the continual exchange of nourishment between workers (these exchanges are a characteristic of all social insects) is the basis of the system. In fact when the contents of the crop are becoming excessively concentrated a large number

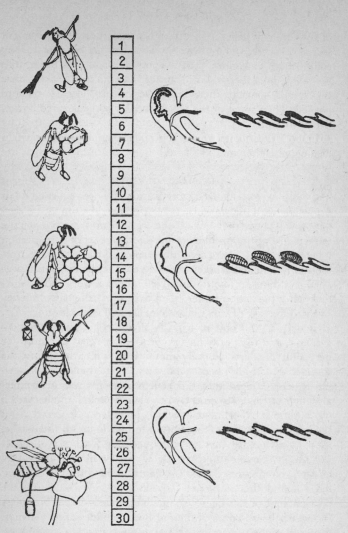

Fig. 1. The development of different activities during the life of an adult bee. The column gives the age in days. The right-hand side shows the development of the cephalic glands, which come into use for the purpose of feeding the larvae.

of bees quickly become aware of it and certain scouts leave in search of water: then, always by means of the bee-dance, they collect more and more of their companions to help them until the whole hive learns, by means of the exchange of food, that the "social concentration" of the crops has been restored to normal. Then some of the water carriers go back to collecting nectar or pollen.

Later I will write about the collection of propolis, when dealing with the work of Lavie.

DIVISION OF WORK

The concept of specialized work has already been touched on in connexion with the water carriers. Moreover the simplest observation will show that not all the workers are engaged on the same task. But the problem of the division of work has caused a lot of ink to flow. There certainly is such a division: for example the newly hatched worker has the cephalic glands well developed and these secrete the royal jelly. Such a worker first of all looks after the young larvae, then from nurse she passes successively to the task of wax maker, cleaner and ventilation worker. It is only at the end of her life that she becomes a honey-gatherer, and then only for two or three weeks at the most, for her adult life lasts scarcely more than a month during the summer. Finally she becomes a guard. Some biologists, following Rösch, thought that this course of events was invariable, although at times it could be reversed. Rösch himself carried out a famous experiment in which hives were deprived of all their young bees (by taking out, in full daylight, all the combs, leaving only empty ones and the queen) or of all their old bees (by simply moving the hive a few tens of metres away: the foragers, coming back from the fields, went to the old site and did not find the new one). Under these new conditions the development of the bees was much modified: in the first case (no young bees), some of the old bees remained in the hive to take care of the young brood; the queen continued to produce, and the cephalic glands of these old bees, which had atrophied, started to work again. In the second case (no old bees), the development of a few young bees was speeded up, their cephalic glands atrophied rapidly; in short they aged quickly, left the hive and by bringing back food saved the colony from starvation. All this is quite true. However Lindauer, from Munich, von Frisch's best pupil, has set right some of Rösch's too rigid contentions. In a hive the proportion of bees that takes on this or that task *depends in reality on the needs of the colony*. The different stages can be telescoped and some bees perhaps may never make wax, for example, or may only feed larvae for a short while, and so on. . . . And yet, at the risk of shocking many prejudices, I must point out something well known to specialists, namely, that a great number of bees in a hive are very busy doing nothing! They sit on the comb without moving, or get into an empty cell and stay there for an hour. Where do we stand now? They prefer urine to dew, they are not so very industrious. What then is left of the virtues attributed to them by the poets? Chastity, perhaps? Well, as we have seen, it is an enforced chastity.

THE LANGUAGE OF BEES

How many disputes and mocking remarks greeted the researches of von Frisch on the dance of the bees, about twenty years ago. Many people thought then that the work was comparatively recent, whereas in fact the first papers on the subject go back to 1926. Besides, von Frisch wrote only in German, and in France few scientists of the new generation can read that. I well remember my surprise when I got some old works of von Frisch from the Museum's central library and found that I was the first to cut the pages. The great word that could not be overlooked was that of *Bienensprache*, language of the bees. Thereupon supercilious linguists (who knew nothing of von Frisch except one or two popular articles of his) started to pontificate and explain to us what a language was and what it was not, and why the bees had no right to speak. It must also be said that what von Frisch had announced was without parallel in all biology: that the scouts can tell their companions the direction and distance of a food source they have just discovered, entirely by means of the rhythm and direction of a special dance they perform on the comb. In the face of such a startling assertion von Frisch's already famous journal of comparative physiology literally fell from the hands of many biologists.

In the end Thorpe of Cambridge decided to get the thing clear. He went one day to Brünnwinkl in the Tyrol, where von Frisch passed his holidays, and without ceremony asked him to demonstrate his famous experiments. "Nothing easier," replied von Frisch. "All you need is an observation hive with a glass wall, a protractor to measure angles, and a watch with a seconds hand. Now I have a saucer of syrup out in the park, on which the bees are feeding. You will be able to find it solely from the information the bees themselves will give you." Thereupon von Frisch left him. Thorpe, as one might imagine, was most mystified. However it does not cost much to make an attempt. Using the watch and the protractor he got a figure of 400 metres distance and 30° to the left of the sun. He set off in that direction. 350 metres, 400 metres . . . He stopped, his heart beating with the stupefaction occasionally experienced by biologists; the syrup was there, he had almost stepped on it.

The Englishman naturally came back enthusiastic. After listening to him several biologists asked themselves whether in the light of Thorpe's confirmation of the experiments it would not be as well to reread the *Zeitschrift für vergleichende Physiologie* more attentively. The battle has been won. Everyone now accepts the experiments made and theories put forward by von Frisch. Better still, some workers, such as Birukow, of Fribourg-in-Brisgau, have found elementary dances in other insects. But since for the last twenty years von Frisch's journal has annually published a number of weighty articles on the bees' dance, it is worth our while to stop a moment and examine this subject.

THE FIRST EXPERIMENT IMPLYING THE TRANSMISSION OF INFORMATION

First of all we put in the centre of a field a von Frisch glass hive which enables us to observe the bees without difficulty; we then mark some of them with spots of paint. This is quite easy to do, as von Frisch has shown, using a variety of colours and, for example, using the head for units, the thorax for tens and the abdomen for hundreds; in this way several hundreds of bees can be marked.

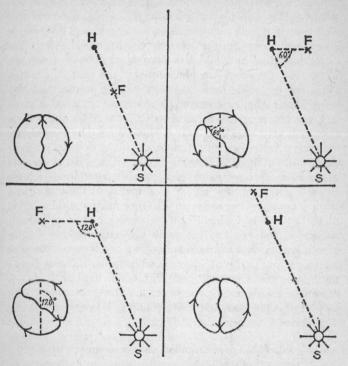

Fig. 2. General plan of the bees' dance, after von Frisch. H the hive: F the food: S the sun. On the left is given the plan of the scout's dance and the angle the axis of the dance makes to the force of gravity, that is the vertical. The arrows shows the head of the dancer and the way she is facing.

Let us now place at the four cardinal points from the hive four saucers of honey at a distance of 800 metres, with an observer near each one. After a while a scout will discover, for instance, the northern saucer; let us suppose she is identified by a white mark on the thorax. She goes back to the hive, and a few minutes later a few bees will come direct to the northern saucer, and not go towards the others, *and white-thorax is not among them*. Thus it is impossible to avoid reaching the conclusion that the scout has by some means or other conveyed to her colleagues the information *where the food is*. Moreover if the saucers are put in the same *direction*, but at different *distances*, it is soon easily seen that the bees only visit the one the scout has discovered; thus, in addition to information about the *direction* of the food the scout has also conveyed the idea of *distance*. What has the scout done and how has she transmitted the information? Another observer, placed by the observation hive, sees her make a curious series of actions, which, though known to us for a very long time, were only interpreted by von Frisch: the scout makes quick runs on the comb in the shape of a figure eight, whilst at the same time she rapidly vibrates her abdomen. The workers surrounding her seem very interested in this action and follow her through all her movements, at the same time tapping the end of her abdomen with their antennae. The longitudinal axis of the figure eight may be more or less inclined in relation to the vertical: the angle this axis makes with the vertical corresponds with another angle made by two lines: one is the line from the hive to the sun and the other is the line from the hive to the source of food. Distance is expressed by the rhythm of the dance: on the whole the slower the dance, the further off is the food, and the relationship is very exact, up to distances of a kilometre, beyond which bees do not often search for food.

But another kind of information also passes during the dance, that of the kind of food. The bees following the movements of the dancer tap her abdomen rapidly with their antennae, as we have already seen; what they are really doing is ascertaining what the floral odour is with which the abdomen is impregnated, for the integument of the bee retains odours very persistently and much better than many other natural materials, such as wood, wool or wax. Thus the information passed, translated into human language, can be put as follows: "Note that at 800 metres, 30° to the right of the sun, there is honey in the colza flowers."

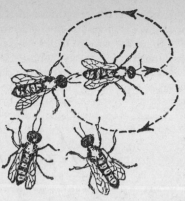

Fig. 3. Detail of the bees' dance, after
von Frisch. three workers start to follow
the movements of the dancer.

DISTANCE AND EFFORT
THE BASIS OF THE "DISTANCE" INFORMATION

If by chance the wind carries the bee forward, the distance
shown by the dance will be less than the true distance, and if
she has to fly into the wind the distance she shows will be greater
than the actual distance, as is easy to see. One might conclude
from this that the indication of distance travelled has less to do
with the distance really covered than with a certain amount of
muscular effort put out to cover it. This is a natural and anthropomorphic interpretation of distance, used among a number of
primitive peoples, who calculate distance in terms of the
difficulties of the journey. Even among the Chinese, not so long
ago, the distance unit was shorter in the mountains than on the
plains, because it took longer to cross the former area than the
flat land.

We must also note that the bees who follow the dancer
gather yet another piece of information from her: the amount
of fuel needed for the journey. The bee is a somewhat uneco-
nomical flying machine and consumes plenty of glucose on a
long journey; a bee thus needs to take on the needed quantity of
fuel in order to avoid a breakdown on the way.

c

As to the explanation of the rhythm which is so exactly related to distance, there is some disagreement among authors. The thing is that there are several different elements in this indication, I was almost going to say "phonemes". When we talk about the variations of the rhythm of the dance we must distinguish between the total time taken to make the figure of

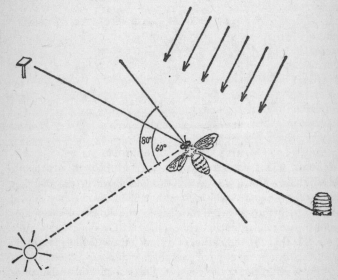

Fig. 4. Displacement due to wind (wind direction shown by parallel arrows, top right) can be compensated by the bee, as the slanting position of the body shows. This means the insect will see the sun at a different angle (80° here) from that of the real angle of the sun and the food (here 60°). Nevertheless the dance showed the theoretical angle of 60°. (After Lindauer)

eight and the quivering produced in the course of the dance. As the frames on cinema film have shown, the frequency of quivering is constant whatever the distance. It cannot then be a measuring factor. The length of time the quivering lasts is in perfect agreement with the distance of the food, as is the length of time of the dance. But the distance covered in quivering does not correspond well with the distance of the food. What is very interesting is the possible intervention of a certain amount of learning here, because bees make fewer errors when they have been able to follow several figures of the dance and not just one; that is to say when they have been able to repeat their lessons several times, not just once; or perhaps they just strike an average of several cycles (because the bee is not a high precision instrument and the length of time of quivering shows certain variations according to the phase of the dance). Heran has carried out research into the bee's method of ascertaining distance when in flight; without doubt the antennae, through which it senses the strength and duration of an air current during flight, will influence the appreciation of distance, in addition to muscular effort. Otto carried out some tests by changing to some extent the intensity of the stimulus in the course of the outward and inward flights: while the bee sipped the syrup in the saucer (when they are feeding they can be treated in all sorts of ways without their taking any notice) he carried it much further away from the hive. Then the length of the return flight was quite different from the length of the outward flight. In this case the distance indicated corresponded to the mean of the two flights: the energy expended in the course of the two flights consequently has some effect on the distance indicated.

INDICATION OF DIRECTION

It should be remarked first of all that the direction indicated is very exact, for the error at 800 metres distance, the distance at which bees habitually hunt for food, is no more than a few degrees. But we have already seen that bees "calculate" (I here use a translation of a term employed by the von Frisch school: *einkalkulieren*) by means of the sun, and this raises a number of difficulties. First of all the sun moves during the course of the day: thus with the saucer of food remaining in the same place the indication of direction varies according to the hour of the day. But if the food source is at a distance of several kilometres (and bees scarcely fly at 20 km. the hour) there can be a noticeable displacement of the sun during a return flight; in this case the indication of direction corresponds to the position of the sun at the time of the dance and not to the position at the moment of taking food. Finally Lindauer, in 1954, discovered a curious phenomenon, that of the persistent dancers (*Dauertänzerinnen*). This happens above all when honey is particularly abundant and the scouts therefore are particularly excited. On these occasions the dancers carry out their evolutions *for several*

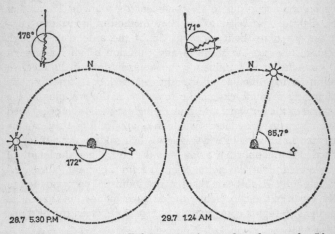

Fig. 5. The phenomenon called *Dauertänzerinnen* or long dances, after Lindauer. The bee was fed at 17.30 to the east of the hive. The creature noted the sun at 172° to right of the experimental table where the syrup was placed. *Top left.* The bee danced with an error of 4°. After midnight the bee is induced to dance again by suddenly exposing her to light. She indicated the supposed position of the sun (85°7′) although obviously she has not been able to see the orb at this moment. The error was only 14° (*top right*).

hours inside the hive, and obviously all this takes place in complete darkness in the heart of the colony, where they cannot see the movement of the sun. Nevertheless the indication they give of direction remains correct. That is to say the inclination given by the "danced eight" changes as the position of the sun changes.

But the sun may be hidden, in which case there are two possibilities: a small patch of blue sky may remain or the sky may be entirely covered. If there is a blue patch the light coming from it is partly polarized and its plane of polarization is in direct relationship with the position of the blue area in regard to the sun. Now von Frisch has shown that the bee can see polarized light; if, for instance, one covers a dancing bee with a polaroid (a special plastic sheet which polarizes light) the poor creature is quite lost as soon as one turns it round. It can be made to dance in any meaningless fashion. Since von Frisch's admirable observation, biologists have found the ability to perceive polarized light in many insects and crustacea. Perhaps this is a special property of the faceted eye, the compound eye, as opposed to that of the vertebrate. Neurophysiologists on their side have inserted micro-electrodes inside the eyes of insects: under these conditions the electric currents that occur when light is thrown on the eye show certain variations according to the plane of polarization of the light. Moreover, following the lines laid down by von Frisch, it has been possible to make a model of a bee's eye with polaroids arranged around a centre; through such an eye a piece of blue sky seems to us, as no doubt to the bee too, to be divided into dark and clear zones, depending on the position of the sun with regard to the observed area.

Nevertheless, if the sky is quite covered how the devil do the bees know where the sun is? At times the clouds are so thick that the orb is completely invisible, at least to human eyes. A curious experiment has its place here, one in which a biological observation contributes to advanced physics, something that used to be very rare but which now is becoming more frequent. Even in thick, cloudy weather, the dancers carry out their evolutions without any trouble, that is unless a light filter, which will stop ultra-violet rays, is put between them and the supposed position of the sun. Thus it must be supposed that ultra-violet rays pass through black clouds and allow bees to orient themselves by the sun. When I told this to Mörikofer, the famous climatologist at Davos, he shouted peremptorily "Impossible! Ultra-violet does not pass through black clouds; either von Frisch is wrong or you have misread him." I did not dare argue much, particularly as we were not discussing experiments of my own. But this is how von Frisch found the clue to the mystery. The Agfa company one day gave him a packet of "Ultrahart" plates, which have an excellent contrast range and are sensitive to ultra-violet light. If a camera lens is trained on a black cloud (not too black nevertheless!) and a short exposure is made, after development a lighter disc can be seen where the sun is. Thus a certain amount of ultra-violet light does come through black clouds, which the physicists never suspected; moreover the bees are much more sensitive to it than was believed. However the clouds are sometimes so thick that the "Ultrahart" plates register nothing. Then the bees stop dancing.

But what happens when the sun is at the zenith? Then obviously the bees cannot calculate their angles, and in fact at such times the workers cease their flights; if at such a moment you take out a few workers and carry them off on a saucer of honey, the dancers on their return are quite disoriented as long as the sun has not dropped 2° 5′ from the zenith. It is a very small angle, nevertheless the bee's eye can appreciate it.

NIGHT DANCES

Some tests made by Lindauer encourage the belief that the bee has a sort of internal clock which, once set going by the sun, continues working for at least the following night. The German experimenter trained a colony to go to two different places. They had to go to the first before sunset, the second an hour after sunrise (bees have a very exact time sense, as we shall see later). Even in the middle of the night one can get the bees to dance by means of certain tricks, the most effective of which is suddenly to shine an electric torch on them through a glass observation panel. *Before midnight the dancers indicated the evening food. After midnight they gave the signal for the morning expedition.* At midnight itself the dancers were completely confused.

HOW DO THEY JUDGE DIRECTION?

During feeding, Otto transported the bees and their saucer of syrup to another place but at the same distance from the hive. On their return to the hive they then indicated an imaginary feeding zone corresponding to the bisector of the angle made by the hive and the old and the new feeding places. However one can give the new place, or the old, an "advantage" by associat-

ing it with some colour which makes its identification easier.
Then the direction indicated leans towards the "advantaged"
side and is not the bisector of the angle. The intensity of the
stimuli can also be modified in other ways, for instance by
shortening the return trip: one simply moves the bees and saucer
nearer to the hive after feeding, but before they fly off. Then the
direction indicated diverges from the bisector of the angle towards the longer arm of the flight. Now most authors have
maintained that only the direction and distance of the outward

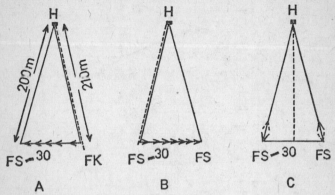

Fig. 6. Otto's figures (1959) showing the importance of the return flight
for bees.
A. A bee from the hive H, feeding at FK, had been moved 30° from there
to FS-30
B. The opposite transfer. A bee feeding at FS-30 has been moved to FS.
The result is the same in both cases; the bees indicate by their dance the
bisector of the angle of 30°, seen in C.

flight are shown by the dance: this is because under normal
conditions the return flight is merely the outward flight in
reverse; when it has to be learnt separately, as in Otto's experiments, it can be seen that the return is as important as the outward journey.

HIGH AND LOW

Von Frisch and his colleagues very much wanted to discover
in the language of bees some indication of high and low
positions. Since bees often go forth to collect nectar from trees

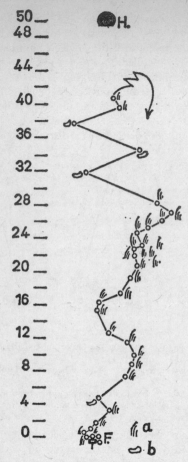

Fig. 7. Plan of odour traces left by a stingless bee (*Trigona postica*), after Lindauer. The small circles show the places the bee rested in the course of the return flight. The creature left odour spots on the stones marked by the figure keyed b in the drawing, or on grass where the spot is marked by the figure keyed a. These odour spots are a kind of Ariadne's thread which indicates very exactly the route to be followed, without using the dance of *Apis mellifica*. H is the hive and F the feeding zone.

in flower it would be natural to suppose that some characteristic of the dance would refer to the height at which the food should be sought. The German workers placed honey at different heights on a wireless aerial: complete failure: the dancers were quite content to show the general direction of the mast, but at the level of the soil, and the bees did not seek honey at the top of the pylon but at its foot. If one puts honey on a cliff strange results can be seen: when the level of the hive is half-way up the cliff and the honey is above it the bees only give the signal "towards the cliff"; and if the syrup is beneath the hive they give the instruction, "Go away from the cliff". It is interesting to note that what our *Apis* cannot do, certain stingless bees (*Melipona*) of America can; these last have a means of communication which may seem more primitive than that possessed by our bees, but which under certain conditions is more efficient. The scout uses an odoriferous Ariadne's thread; that is to say, in returning from an expedition she stops from time to time to nibble a stone, twig, etc., and the mandibulary glands leave a spot of highly odoriferous substance there. Back at the nest she just emits a special buzz which draws the attention of her colleagues to herself. They then leave the nest in a crowd and, following the "Ariadne's thread", find the food in no time. Thus the Melipone are able to find the honey on the top of a pylon and to tell their colony about it, simply by using this system, leaving an odoriferous spot from place to place.

THE INNATE AND THE ACQUIRED

All biologists agree that Lindauer almost equals von Frisch in the ingenuity of his experiments, but in my opinion he surpassed himself with his experiments on innate and acquired characteristics in the dance of the bees. He raised a colony in a dark room; they hatched there and lived there without ever having seen the light of day. Finally he let them out about midday and trained them to take syrup from a saucer situated to the south; then the bees were returned to the darkness and not let out until noon of the following day. They were then seen to go towards the south. After that they were moved to a new position and only allowed out *in the early morning*; they showed great confusion and many bees went towards the east. This must have been due to the fact that these bees had never been out in the morning, had never seen the sun except to the right and were not capable of calculating (*einkalkulieren*) the displacement of the sun. Apparently leaving them outside for a whole afternoon is not enough to get them to the stage of "calculating" their position. But if they are left outside for *five more afternoons* they then learn to adjust to the movement of the sun *during the morning*: always provided that it is worth their while, because if one merely lets them out and offers no bait the apprenticeship takes place much more slowly. Lindauer has expressed the matter quite clearly: what is innate is the bee's ability to use the sun as a point of reference when orienting herself, and from the very first flight too; consequently bees have an innate facility to associate the sun with their movements during the day, *but they have to learn how to do it*. Thus the night dancers, about whom we were speaking above, must "calculate" the position of the sun as a result of their daily experiences. Becker also reports that young bees cannot find the hive as easily as old bees, and can only return to the hive from short distances. Their performance is worse as regards distance than direction.

IS THERE ANY DEVELOPMENT IN THE LANGUAGE OF THE BEES?

The biologist Kalmus suggested that this was the case. Once when he was in South America he transferred a colony from the northern hemisphere to the southern and it found itself quite at a loss as regards direction; nevertheless bees brought from Portugal 400 years ago are able to orient themselves perfectly well. Thus, Kalmus maintains, there must be something of the innate in this "knowledge" of the route followed by the sun, and consequently some sort of evolution of the bee's behaviour has come about during the 400 years since the European conquerors first brought *Apis* bees from the northern hemisphere. But Lindauer does not agree: according to him, Kalmus's observations were not conducted over a sufficiently long period. If the transferred bees are watched over a longer time, though there is great confusion at first, after four or five weeks, when the young bees who have never seen the sun in the northern hemisphere hatch out, everything returns to normal.

DIALECTS

Here is something that will disturb the linguists. How can such a word be used about bees? What nonsense! But let us look a little more closely: all races of bees (there are five or six in our

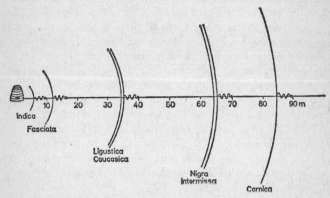

Fig. 8. Racial differences shown by the points at which the circular dance (a simple announcement of the presence of food near by, with no indication of its direction) becomes the great figure of eight directional dance. In each race or species, within a certain distance bounded by an arc, only the round dance is given, outside of this arc the directional figure of eight is performed. (After Lindauer)

climate) always limit themselves to dancing in a circle when the food is near. Now the critical distance, the distance at which the figure eight dance replaces the circle, varies according to the race. Moreover the tempo of the dance varies: the Carniolans dance quickest, being followed at some distance by the Germans, the Telliens, and then by the Italians and the Caucasians, which are particularly slow. It is quite possible, by means of bee-keeping techniques, to mix races in the same hive, and mistakes are then noted. The bees cannot "understand" one another, for a Carniolan who follows the dance instruction of a Caucasian scout does not know that, to indicate the same distance, this race dances more slowly than do her compatriots.

DIALOGUE

Now, the linguists will say, you have gone too far. They will then give us the definition of dialogue in human terms and affirm what many are quite convinced of: that the dance only indicates where the nectar is. This is by no means the case, as we shall soon see when we examine the exchange of information after swarming. A little before the swarm leaves the hive scouts go out to look not only for food but also for shelter. Now Lindauer wanted to know what guided them in their choice, so he took on to a barren Baltic island a number of apparently suitable shelters and several hives of bees which were ready to swarm. It was then easily seen that the swarms preferred old skeps to wooden hives, a place sheltered from the wind and one near the site of the original hive. The site had also to be in the shade and free from ants. The swarm then left the hive and settled in a cluster on a near-by branch. On its surface were the scouts who wanted to direct the swarm to the place they had discovered; there they carried out their famous dances, continuing for days at a time if the need arose, and always allowing for the movement of the sun.

But Lindauer, who knew a great deal about bees, indulged in the final cruelty of showing two scouts two different sites, distant from each other, but otherwise equally suitable. The two scouts' dances indicated opposite directions. If one of them was less obstinate than the other, she soon stopped dancing, went to see what her companion was doing and little by little fell in with her plan. But if they were both equally obstinate, the swarm would be split into two halves, each going off with one of the scouts to her chosen site. Actually, the misfortune was not irreparable since there was but one queen so that one of the halves was necessarily queenless, and this half soon joined the other in its new home.

COMMUNICATION WITH MAN

Since we know the bees' language so well why not use it to communicate with them? Obviously this may seem a foolish idea, more suitable to the realm of science fiction. Nevertheless Steche put it into practice. It must be noted that, contrary to

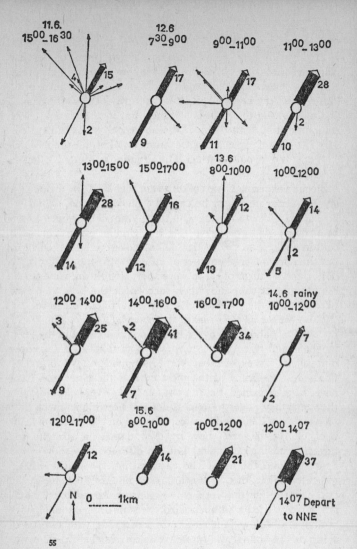

55

Fig. 9. A little after swarming it may be found that two separate groups of scouts have found two possible sites for the swarm, both equally attractive, and such scouts, in dancing, can show two quite different directions. Usually one of the two groups is "converted" to the direction indicated by the other. The number of bees dancing is shown by the thickness of the black arrows below the time figures (after Lindauer). It will be seen that the competitive dances can last for many hours, even days.

what is usually believed, bees do not "know" man. In the first place they are too short-sighted to see him properly and probably he appears to them as an enormous and indistinct mass that smells very badly. Apiculture is a way of exploiting bees, and of knowing their needs at times, but is never a matter of communicating with them. If the beekeeper is not stung, or is very little stung, it is solely because he takes the proper precautions and avoids getting into situations where a sting is inevitable.

Steche's experiment was of quite another kind: by means of a low-frequency generator he caused a piece of wood, or a dead bee, inside a hive to vibrate with the rhythm of the quiver in the bees' dance. He even gave it a semicircular movement at a certain angle from the vertical. The bees were interested in it, but then flew off in a number of different directions. But if the object and a saucer of syrup were scented with the same smell, the attempt at communication succeeded; the bees then flew like arrows in the direction given by the artificial creature.

However *one mystery remains: the showing of a detour.* The lines I have written above are a very abbreviated account of the bees' language and it would be quite wrong to think that all is now quite clear. The experiments von Frisch made on the subject of the detour are some of the most curious there are and have never been explained. Bees do not like flying over hills, although they can; they prefer to make a detour. Now if one puts a hive on one side of a hill and a saucer of syrup on the other the scout bees, on their return, show by their dance the straight line direction (as the crow flies) but the distance they show is the detour (round the hill). The workers find the place with no difficulty. What, then, is this unknown nuance they give to the dance which tells the workers they must make a detour? Bisetsky, in 1957, tried a number of experiments using a tunnel, with a number of elbows in it, put in front of a hive, with a saucer of syrup at the end. If the tunnel was straight the direction was shown quite normally in the dance, but if there was a right angle in it the bees signalled (as in the case of the hill) the true, as the crow flies, direction, and the actual distance flown.

Thus the question remains unanswered; and in spite of

numerous papers on the dances there remain, without doubt, still as many problems to be solved: the well of knowledge is deep and inexhaustible.

MY PERSONAL ADVENTURES WITH BEES

I took over the directorship of the bee laboratory in 1949 ... a far-away beginning, not without charm, as is everything at the beginning. The country had been badly set back by the war and was accustoming itself slowly to peace. We were among the few who knew that a scientific renaissance, such as France had never seen before, was about to start. I was there, dreaming, amidst the cheerful bustle of my pupils, who were unpacking apparatus, and of workmen hammering, sawing and planing all over the place; I dreamt of the past; of the dreary period between the wars which followed the terrible defeat of 1914–18, for a war which is won by millions of deaths cannot in any case be called a victory. I saw again the almost empty laboratories where I had been shaped for well or ill, without equipment or credit or aid of any sort.

It was a triumph of string and sealing wax, old tins and the flea market, thanks to which, at the cost of a great deal of time, we nevertheless did obtain the apparatus we needed and could not buy. French science was going downhill at a tremendous rate and was not even conscious of the fact, turned, as it was, towards the glories of the past, like all the rest of the nation. We had been powerful, admired and wise; it seemed too unjust that we were no longer so. The only resource left was to refuse to face reality, and everyone was doing that.

What difficulties we had to overcome, how many bitter draughts we had to swallow, we handful of young workers animated by the sacred fire and determined to follow science come hell or high water! No one nowadays can imagine it; we were mad, happily, mad with the thirst for knowledge, and we did not even take notice of innumerable difficulties awaiting us. To quote the ironical words of one of our old teachers it was "selection by hunger". It went on for twenty years, under the gloomy conviction that a scientist must necessarily be ascetic, unmarried and above all childless. That was what the directors of research thought then, and there are a few still left today. When I remember these events I can still hardly control my fury. There was only one thing to be done: think only of science, of the great mysteries that surround us, of the "calm courts beloved of the wise", *suave mari magno turbuntibus aequora ventis* . . .

But I must pull myself together. All that is over. I now have my own laboratory. Now I can work and show my worth. At first sight things were bad. A villa, more or less gutted, to be transformed into a temple of science, five or six rotting hives in the park alongside and a number of pupils, willing, but knowing little of bees (their chief didn't know much either, for that matter). Well, at least, we shouldn't have preconceived ideas! But all the same, when one remembered that the laboratory at Munich had been running for twenty years, under the able direction of von Frisch, a laboratory solely concerned with bees, and that more progress in apiculture had been made in those years than in the preceding ten centuries: and that we should have more or less to keep up with people like that. . . . But then those are the things that give value to any achievement.

Before starting to work I had to decide on my general plan of action. It was this I was thinking of in the spring of 1949 as the bees started to forage here and there. Everybody told me that as I now had a well-equipped laboratory I should certainly study the dance of the bees, as did von Frisch; I replied that I certainly would not. How could I catch up with that wonderful man, in a field he knew far better than I did? One had to find something he had not thought of, or had not had time to follow up. We neglect far too much these questions of policy; I feel it more and more when I find France exhausting herself to make space rockets at a time when the Russians and the Americans are at least a light-year in advance. Should one be discouraged? Not in the least; one should overcome it by the power of thought alone; one should find parallel or divergent lines. These are infinite as knowledge itself is infinite.

And now I had carefully to consider my new subjects. There they were in front of me in an observation hive, leaving and entering through a tube passing through the observation window. How strange that was! I had studied many migratory insects, such as swarming locusts, which are very gregarious but not social; I had also worked a little with cockroaches, crickets and grasshoppers. But I had never seen anything like these bees. It was a dense mass of brownish animal life, moving gently and continuously, like particles in suspension seen under a microscope, and showing "brownian" movement; and it was hot, as could be felt when the back of one's hand was placed against the glass, at the centre of the mass. It is here that the queen is often found and that the temperature can reach some 30°C or more. When the insects are separated, by shaking the container, they start joining up again, just as a magnet draws iron filings together. Apparently bees like being side by side. Well, what would happen if one isolated them? Here was the first possible investigation. Then, what is it that draws them one to the other? Second subject for research. On the other hand, they were coming and going freely for I know not what purposes outside any control; what would happen if I curtailed their liberty, for instance by making them forage inside a box? In this way I would control every factor in the experiment, the essential law in science.

What else could I do? I had decided against the bees' dances and sensory physiology, as the Germans far surpassed us in both these spheres. I had a confused feeling that the main characteristic that distinguished the social insect was its life in a society. A blinding glimpse of the obvious, you may say. Not at all, if you think about ways and means of studying the influence of the group on the individual. This is far from simple; but Hess had published in 1942 a series of observations and some strange experiments in which he had shown that the presence of the queen prevented the development of the workers' ovaries. For the workers are female; they have ovaries which are atrophied and non-functional. But one individual, the queen, has fully and enormously developed sex glands. If she is removed from the hive, the workers' ovaries begin to grow, and if she is put back they get small again. This was a strange social reaction, with only one paper published on it up to that moment. Here was another subject. And there were still two more. I had been fascinated by the balls of multicoloured pollen which collect on the hind legs of foraging bees and are their sole source of nitrogen; very little was known at that time about pollen collection: we could study that. Moreover we should not forget that we were the Institute of Applied Research; we had to think of the urgent practical problems of the apicultural industry, even if it obliged us to leave for a while the intoxicating realm of pure science: for instance, devastating diseases may often lay waste a hive, among them being a particularly severe one, acariosis ("Isle of Wight" or "acarine" disease). A small mite, *Acarapis woodi*, gets into the respiratory channels of the bee, breeds in the big respiratory ducts of the prothorax, and gradually suffocates the insect. Hardly any cure was known and the trouble spread every day: consequently I gave this subject to one of my pupils to study and a strange thing happened to him. This I shall tell. in the next section and it well illustrates the unexpected byways that lead to a discovery.

COMMON DEFENCE AGAINST DISEASES IN THE HIVE

Lavie started by infecting some bees with Isle of Wight disease, which, unfortunately, was not very difficult, and he tried different kinds of gases with the object of killing the mites in the tracheal of the bees without killing the host. A dangerous task, and one which at first did not have much success. I remember one evening I was examining the complete run of the *Archiv für Bienenkunde* which we had just acquired. This apicultural bible contains some real treasures. The laboratory around me was beginning to come to life; everyone was busy. The happy young voices of my assistants resounded as they bustled about because, though I myself was approaching forty, the average age of my colleagues was well below thirty. At that moment Lavie came down from the minute room he had somehow managed to make into a laboratory and said, "Sir, I really think that we have found the answer at last; I have carried out a fumigation with sulphur dioxide and this time, believe it or not, the mites are dead and not the bees." Delighted, we both went up to the first floor. I checked with the microscope: there was no doubt that the mites were dead, really dead. Victory! But let us have a look at the control bees, which had been in similar chambers, under the same conditions, except, of course, exposure to sulphur dioxide gas; amazement: their mites were dead too! We had never seen such a thing: we had never seen *dead* mites in the bees' tracheae. Incomprehensible. Ah well, it was just one of those experiments that fail for some unknown reason, as so often happens, unfortunately, in biology. I went down, leaving poor Lavie most disconsolate. Next day he came to see me again, rather excited: "I have looked into the matter of those bees with the dead mites in the tracheae more thoroughly," he said. "We must take note of the fact that they all come from the Seine-et-Oise. Moreover, those mites are surrounded by some round irregular bodies in fair numbers. Can it be that the mites are ill or have been attacked by some natural enemy?"

It was by no means a ridiculous suggestion. Man would long ago have vanished from the earth if the innumerable pests which eat his growing crops did not themselves have enemies. One could easily suppose that the mite also had enemies, for, if not, it would no doubt long ago have extinguished all hives.

Nevertheless, no one had ever reported such a thing, and "unlucky are the facts that do not fit the theory!" as Eddington jokingly said. Under the microscope I too saw the corpuscles that Lavie had found; they looked like nothing I had ever seen, except perhaps yeasts.

"Should one perhaps try cultivating them," suggested Lavie, "and then infect bees suffering from Isle of Wight disease?"

I agreed, but without much enthusiasm. However, success in part crowned our efforts; the yeast—for it was a yeast, a sub-species of a known yeast—grew very well. We atomized an attacked hive with it and often got a great improvement and sometimes a complete cure—which, by the way, certain veterinary circles, which had no great reputation for intelligence among research workers, would never admit. But the ones we had to deal with then were supremely stupid! Animal disease cannot be looked after properly unless one has a degree in veterinary science, even if you are a D.Sc. It is better for bees to die within the law than be cured illegally. It was no good quoting the case of the great Pasteur, who committed the same crime, though far more gloriously than we. However, let us forget these mediocrities.

For the story was only beginning. I had been impressed by Lavie's success in the treatment of the acarine (Isle of Wight) disease; nevertheless I felt there was a need for a more thorough investigation. For example, if one sprays a swarm with a culture of these yeasts, how does it get into the tracheae? We are dealing with relatively immobile organisms. I then suggested to Lavie that he try making cultures every forty-eight hours, by pushing a platinum wire into the furry covering of the bees. We could then see if it were possible to recover the yeast. No sooner said than done. A few days later Lavie came to see me in some perplexity; he had made cultures in about 100 tubes and had found the yeast in a few of them. Only nothing else grew in the tube, or very little . . . frankly I did not believe him: after all, everyone knows that the hairs on a fly's body carry a lot of dust and germs of every kind. Bees are very furry, and they brush up against all sorts of things more or less dusty and dirty. There seemed to be no reason why cultures taken from their fur should not be successful. What was to be done then? Start again, that is the law taught by experience, using more tubes and taking greater precautions. That we did; and once again almost nothing grew.

This was too much! Perhaps our culture medium was exhausted, though we couldn't imagine what had happened to alter it. We threw it away and made some new medium, and Lavie started chasing flies which were all over the laboratory windows—there were only too many of them. We scratched the back of these flies with the platinum seeding wire and started another series of cultures which grew this time: all the culture tubes were full of a great variety of colonies of fungi and bacteria. But if we used the same technique and medium with bees, nothing grew. A whole bee, killed by the cold, put into the middle of a plate of medium, only allowed a few microorganisms to grow, or even kept the culture medium sterile.

Astonished, we looked at our tubes. I felt a tremor of excitement run down my spine. Sometimes wrong, often right, I find that in my life as a research worker I have a premonition from

time to time that something important is going to happen. I
then become filled with an expectant anguish, with such impa-
tience that I am tempted to take short cuts in the experiments in
the hope of getting at the truth earlier; so much so that often I
have to start it all again. It is quite impossible that the fuzz on
bees does not contain as many germs as that on flies, and if one
could not make them grow, then either *they were dead* or else they
were inactivated, doubtless by the action of an antibiotic secreted by the bees. This is not as surprising a supposition as it
may seem to be; on the skin of man and in his external secretions
an antibiotic, *lyzozyme*, is found, which kills benign germs falling
on our skins in millions. It is perhaps for this reason that they
remain benign; they cannot really get at us.

An intriguing hypothesis, but one still needing to be proved. I
thought of extracting some bees with hot alcohol, which dissolves a considerable number of substances, then evaporating
the alcohol and mixing the residue with a culture medium on
which we would sow the current range of bacteria. This was
quickly done, but we had to wait for one or two days for the
cultures to germinate and grow, and I was mad with impat-
ience. At dawn on the third day there was no doubt about it; a
great deal of growth in the control (containing none of the
presumed antibiotic) tubes and nothing in the tubes containing
the bee extract. We were very happy that day: it was one of the
moments that repay with interest the thousands of failures;
this is why science is an exploration of the most absorbing
interest and how, as has been said, it is one of the few human
undertakings that have succeeded.

Something else was floating around in my head; I vaguely
remembered, as did Lavie, some very old reference, about 1907,
that said something similar. Our library hardly existed; but it
was relatively easy to find the paper in question. It was by a
certain Mr White, a bacteriologist, who had the idea of studying the internal flora of the hive. He also started from the
assumption that bees must bring countless quantities of germs
into the hive, and that it would be interesting to study those
that could survive there. He too pushed his platinum wire over
the wax of the combs and then sowed numerous tubes containing

culture media. *Almost nothing grew*: a fungus, a yeast, a bacterium. This greatly surprised White, who had no idea what to make of his results, which, of course, were soon forgotten. It was too far away from the era of antibiotics. However the question arose, if nothing grew on the combs was there an antibiotic there too? There was indeed, as an experiment with an alcoholic extract of comb showed us three days later.

We were now getting to a particularly interesting stage: the formation of a general hypothesis. An antibiotic on the bodies of bees, another on the comb; we soon found a third, guided by a number of old scientific papers, in pollen, a fourth in royal jelly (this last has even been isolated in the pure state), a fifth in honey, a sixth in propolis, the gum that bees collect from the buds of poplars and other trees and use to block up cracks in the hive. The last, the one in propolis, is one of the most powerful and has another characteristic too, it is *antifungal*: no mould ever attacks propolis. It is also *antigerminative*, as Lavie found. If you put potatoes or wheat grains into a hive and then try to germinate them you will not succeed. Is this the reason that the millions of pollen grains stored in the hive by bees do not germinate? Because if they did, the increase in volume would quickly spoil the alignment of the comb.

As we shall see later, the hive is undoubtedly a true organism, not just metaphorically one, and an individual bee is but a cell, with no importance as an individual. And like all organisms, our own body for instance, it protects itself against infection. This must be the case. We have seen above (p. 22) what large quantities of food the hive needs in order to live; and all botanists know well that flower pollen is saturated with quantities of fungus spores, yeasts and bacteria. Honey contains quite an appreciable quantity. Consequently if all those organisms were not *inactivated* by some means or other the hive would turn into a rotting mass. Conditions there are so favourable: a high temperature (33–40°C) and a high humidity. Yet nothing is cleaner than a healthy hive; it smells of nothing except wax (and that smell comes more from propolis in fact); all debris is carried outside and the worker bees never defecate in the hive, but always outside.

No doubt, one could reply, there are fascinating resemblances between a hive and an organism, but the social function of antibiotics, curious as it may be, does not push the analogy very far; true, but we have other arguments.

DEATH WHEN ISOLATED

I have long been struck by an evident truism: what characterizes the social insects is that they are social, that they do not live in the solitary state! What would happen if they were forced to be solitary? At the time this idea struck me I was working in the laboratory of my master Grassé, who was amused at the idea; it was so simple that no one had ever thought of it. We then isolated bees, ants, wasps and termites, all giving very different results. Wasps did not get on so badly, neither better nor worse, as far as one could see, whereas bees, ants and termites apparently all died at the end of a few days. Surprised at the result, but not knowing what to make of it, we thought that some day or other we must study the matter more thoroughly. What, then, is this social influence which is so powerful that the insect cannot live long without it?

It was not for several years that I was able to take up this matter again. Isolated bees definitely died more quickly than groups of bees; but groups of two bees showed a survival rate a little above that of isolated bees. The group had to be about 40 bees in a volume of two or three hundred cubic centimetres to get a more or less normal survival rate. Moreover if the isolated bees were kept without food and water, but separated from a mass of bees only by a metal screen, they lived; they could be seen exchanging something through the meshes of the screen by means of their mouths. Such exchanges are continuous among the bee people and in all social insects. It is thus probable that survival is due to some substance that only groups of bees know how to make and that the others are unable to get. What sort of substance is it? I then fed certain isolated bees sugar syrup only; to others syrup plus casein, the protein base of all artificial diets; to yet others casein supplemented with different vitamins, and to a final group vitamins only. This last treatment, and this only, gave lives as long as those of normal bees in a group (especially when thiamine or biotin was added). It was a simple enough experiment, but how were we to interpret it?

As is often the case, it can only be interpreted properly by someone whose subconscious mind is, as it were, strewn with vast slabs of reading; someone who knows of his subject very nearly all there is to know, so that apparently forgotten experiments are suddenly remembered; someone who has become what the good people call a *savant* (we may note in passing that such a word is never used in laboratories, except ironically). This is the end we are all trying for, but hardly ever achieve. But at last, after ten or twenty years at the job, many of the pieces of the jigsaw puzzle end up by putting themselves into place. I remembered, for instance, that Haydak worked on the analysing of young and old bees for certain materials. He found that the basic chemical composition was very different. In particular young bees are much richer in vitamins than old bees. One might thus postulate that the old suffer from a chronic loss of vitamins, and that they have to make these constant buccal exchanges with young bees in order to get supplies. It was a neat and satisfying explanation.

Unfortunately it is not true. For if so the young isolated bees should live longer than the old foragers. This is not the case; or rather the difference is not enough to enable one to interpret it that way. What other explanation is there? That isolation favours the loss of vitamins? But what does loss mean here, since bees do not defecate in the cages and in reality lose nothing? Here again is the mystery, the brick wall across the road, that one so often finds in biology. We must find a way round it then.

THE ROLE OF THE QUEEN

There is in a hive one quite different individual which has, very improperly and for a long time, been called the *queen*; she gives no orders and appears only indirectly to control the activities of the hive. She is a reproductive organ, an ovary, which not only produces eggs but also a certain number of hormones; these gradually regulate a great number of social activities in the hive.

It is not a long time, hardly more than a few years, since our ideas on the part played by the queen have become clearer. It is a strange story, full of meanderings and wrong roads, as is all research away from the beaten path.

I gave Mlle Pain the task of studying the factors influencing ovarian development in the workers. We had noted that, in effect, the ovaries were atrophied in general, but that their development could be much modified. For instance, if the queen is present, the workers' ovaries are small; but if she is not there the ovaries will grow to the size where eggs can be laid. Why?

Why indeed? And where do we start with such baffling facts? Of course, where we usually start in science when faced with a strange phenomenon: that is to say, by the humdrum examination of circumstances surrounding the subject, such as environmental conditions—temperature, for example (not of great interest as bees almost always live at around 30°C), age, food, and naturally isolation or social living. Age, for instance: however atrophied the ovaries may be they do show a certain development cycle, as may easily be seen in a hive deprived of its queen; they grow until about the tenth day, when eggs appear and increase, but only if the workers are given a nitrogenous food, for if only sugar and water are provided the ovaries remain very small and do not develop. The kind of nitrogenous food that suits them best is found in pollen. Another very strange phenomenon is that the bees also need the presence of their sister workers if the ovaries are to reach a proper size. In the case of isolated bees the ovaries are considerably retarded even if nitrogen is given in profusion. And still stranger, bees in groups of two can overcome the delay in development, but the ovaries of one grow much faster than those of the other, as if one bee were donor and the other receiver of the favourable substance, or as if the digestion and use of protein material could only be successful in a group. We thus come back to the mystery of the *group effect*, which is the central theme of the social physiology of bees and, without doubt, of all insect societies.

Now once Mlle Pain knew all the facts as regards the development of the ovaries she could attack the main problem, the part played by the queen. Fortunately, the problem is clear; it is the enormous effect the queen has: as soon as she arrives the ovaries of the workers are reduced to thin, scarcely visible threads.

At this point something new happened. Mlle Pain introduced an old queen, who died the following day, into the cage. The bees continued to pay attention to her and to tap the body with their antennae; there is nothing very surprising about that, for beekeepers have been aware for a long time that even a dead queen attracts the workers. One has to know how to approach a problem and we were on our toes that day. All we had to do was leave the body there and *see what happened*; and after a few days the ovaries were as small as they would have been had the queen been living. The necessary conclusion was that the inhibition must be of a chemical nature and due to a substance sufficiently stable to persist for a certain time in the body of a dead queen. Moreover pieces of a dead queen, cut up with scissors so as to make her unrecognizable, another queen reduced to powder put into a silk sock, and even an old queen from a collection three years old all gave the same result. I still remember our bewilderment at the results of the last experiment with the old queen from a collection. Refusing to believe her eyes, Mlle Pain set out to find queens from dusty old boxes and to try them; she always got the same result (though with certain fairly important differences, according to the origin of the specimens). Thus, the inhibiting substance must be of unbelievable stability to be capable of persisting on old dead bodies kept in such a crude way in insect collections.

This was an interesting conclusion and one that would have sufficed twenty or thirty years ago. We have improved on that nowadays, we want to know the chemical formula of the substance. Once such a thing could not be dreamed of, because analysis needed a big sample before it could be undertaken: remember that there is but one queen per hive and that the active material cannot be more than a fraction of a milligram. But by means of a new apparatus used by the chemists—the gas chromatograph—we can now work even with such minute quantities. To understand how it works we must give a little thought to chromatography. If, for instance, you let a drop of ink fall on a piece of blotting paper you will see not just a dark ink spot but a series of concentric zones, alternately lighter and darker. The different pigments in the ink distribute themselves, due to the phenomenon of absorption, into distinct zones, where they are found almost in the pure state. You can do the same thing with a column of aluminium oxide, for instance; if you pour ink in at the top of the column you will find the same zones from top to bottom, and you merely have to cut up the column and wash them with a special solvent (the eluant) to get an almost pure substance. This chromatographic absorption has innumerable uses and many improvements in the method have been made. The recent invention of the gas chromatograph represents a considerable advance. It is only used for fatty substances, which are vaporized in a very hot current of argon. The gas is then passed through a column of an inert substance which selectively retains the fatty acids in distinct zones. If the passage of argon is continued the zones leave the column in a predetermined and constant order, and one which is characteristic of the fatty acid in question. If the current of gas carrying them is strongly ionized by means of radio-strontium and passed between two electrodes, the current passing is more or less according to the nature of the substance carried in the gas stream. An apparatus is arranged to draw a curve, the reading of which shows not only the nature of the substance but also its proportion in the mixture.

The apparatus is not very impressive; it is a big, black, metal box with a few suitable buttons and glowing bulbs. It regularly spills out a band of paper on which a moving pen slowly gives us, in the form of a curve, all the information we need. All one need do is to put a few milligrams of the crude substance into the apparatus. Many times I have looked at this almost miraculous machine, thinking of the stupendous technical progress that has been made, a progress which is ever quicker, as indeed is that of science itself. I remember reading a science fiction novel in which the hero, having landed on some unknown planet, put samples of the flora and fauna into the "synthesis analyser" which soon afterwards gave the proportions and formulae of all the ingredients! We have no such apparatus, but perhaps it is not far off. What will revolutionize chemistry is biology.

. Even though we needed only a few milligrams to put into the chromatograph we did not have them. We then appealed to all the beekeepers in France and Navarre, who periodically change their queens (egg-laying drops off at the end of two or three years). We also asked the big American apiculturists, some of whom have more than 20,000 colonies. All went so well that we soon found ourselves with more than three kilos of queens, a fantastic quantity which no doubt no one had ever had before. The chemists could now start the first steps of the necessary preliminary purification before putting the material into the gas chromatograph.

Mlle Pain soon found that a few drops of the extract put on a piece of paper had a powerful attraction for the workers and inhibited the production of ovaries as much as a living queen; still further proof of the chemical nature of the inhibition process. But as the purification process went on we had more and more troubles. The active material seemed to comprise two different groups of substances, one of which was much more volatile than the other; it was also unstable, so much so that it played one or two tricks on us as we altered the technique of the operation.

At this time Butler, of Rothamsted, was working on a similar theme: that when the queen is present the workers do not raise up replacement queens (except under special circumstances, which are discussed below); that is to say they do not build the large, round cells called the queen cells round the young larvae. But if the queen is removed the building of queen cells starts almost at once. Butler noted pretty quickly that extracts from the queen had the same effect, and he was fortunate enough to isolate the active principle from the body of the queen. It has the resounding name of 9-oxydecanoic acid, a description which is crystal clear to a chemist. Pleased with his success, and that is easily understood, the impetuous Englishman hurried to trumpet to the world that he had discovered the royal hormone, the substance which regulates, as if by magic, all the hive's activities.

At Bures we received Butler's papers; while rejoicing at his discovery we could not help thinking him a little rash. "The royal hormone"? It would have been better to put this word in the plural! For in the interval things had become considerably more complicated.

To understand this let us go back a bit. Truly the queen is a remarkable insect, who induces a whole series of reactions in the workers (we still do not know them all!): (1) she attracts them as a magnet does iron filings; (2) she stops them making queen cells; (3) she inhibits the development of their ovaries; (4) by contrast she induces them to make many combs with cells for workers and males. Now Butler had only studied reaction No 2; whereas we were interested at least as much in the other three. And it is there that things became complicated. Let us deal with attraction first of all. The royal substance isolated by Butler, 9-oxydecanoic acid, attracts nothing at all; bees are quite indifferent to it. To become attractive it must be mixed with another group of more volatile substances, also extracted from the queen, and no more attractive than 9-oxydecanoic acid. . . . *But the mixture of these two inactive substances gives an active one.*

Here is something that may surprise the layman! But there are other examples of this paradox. I suppose that you, like me, enjoy coffee, always provided it be good. I am not speaking here of those horrible mixtures, sold powdered, whose smell is an insult to the nose and whose taste is a violation to the tongue's taste buds; coffee should persuade not bully. No, by coffee I mean those subtle mixtures of well roasted grains, whose aroma decongests the brain and leads it to poetry, science and all the arts. God knows how many chemists have tried to reconstruct that aroma. It depends on a strictly proportioned mixture of some fifty substances, each of which, taken separately, has no, or very little, odour of coffee. An error in the proportions of one or the other of these substances upsets the mixture; *it no longer draws coffee lovers but becomes repulsive.* Thus substances having very different characteristics from those of their constituents are often found in nature.

Butler did not notice the inactivity of his *queen-substance* because he gave it mixed with honey, which the bees like any-way; they took in the queen-substance at the same time as the honey, and, as a result of a whole series of physiological reactions, still somewhat mysterious, became incapable of building queen cells. But Mlle Pain put the materials she was studying on strips of paper on the floors of the experimental cages she was using: any reaction by the bees must then be due to the substances themselves. When it was a crude queen extract the workers were drawn to it. They feverishly tapped the paper with their antennae, they licked it, they rubbed their abdomens on it and ended by cutting it up into little bits. There was thus a whole series of reactions only produced by the queen extract. This method of sensory excitement is indispensable; if in fact the substance is mixed with food it may well stop the production of queen cells, but it does not prevent the growth of ovaries.

This can be explained in a few sentences, but it took us thirteen years to understand. God knows how many stupid hypotheses we formulated before realizing we were the plaything of a group of substances none of which could work alone.

Now, as I wrote before, the story of the queen hormone still needs a chapter or two to make it complete. It is hardly more than a few months since I realized that our queen extract might attract plenty of bees, stop the growth of ovaries and the construction of queen cells, as much as you like: but there was another reaction, very characteristic of the presence of a queen, *the building of wax comb, and this we had never seen when our extracts were present*. There was only one inference to be made: our method of extraction with boiling alcohol was insufficient, or, more probably, it destroyed the substances responsible for the reaction. It was also a question of a basic chemical phenomenon because a dead queen will produce the comb-building reaction. I will not dwell on the different recipes we used, but it took us several months to extract from the queen, without destroying it, the precious substance giving the construction reaction. It is a volatile and very unstable material; the queens must be extracted in a cold room with a mixture of ether and acetone at 0°C, then the solution must be gently evaporated by a current of cold air. This gives a few milligrams of a white fatty substance having a faint smell, which, put on a piece of paper, strongly attracts bees with no queen and also induces them to build beautiful wax combs.

But that is not the whole story, as we well know. Why, for example, do the bees at certain moments set to work to build the large male cells instead of the small worker cells? Does it arise from some subtle modification of the queen's hormone output, or from something else? We know nothing about this. And the males themselves, traditionally considered as idle drones, of no use to the hive, is it unthinkable that they too might emit hormones? As we have no experiments on these points no reply is possible. Finally perhaps the workers are not inert, passively submitting to the hormonal dictates of the queen; what could the mechanism of their reciprocal interactions be? On this last point we do have a few recent facts in addition to the work of von Frisch on the dances.

THE COHESION AND DEFENCE OF THE COLONY

We cannot too often repeat this salutary truism: "What characterizes social insects is that they live together." But why? Could it not be that they are drawn one to the other? We can unravel this tangle, thanks to a very simple experiment Lecomte made in my laboratory some years ago. Put a handful of bees without a queen in an empty, dark, closed box and have a look at them from time to time: after about ten minutes the scattered workers will have formed little groups: wait a little more and they will all be in one group "having themselves a ball", as might be said. It never fails, and always happens the same way and is most striking. Everyone has seen it, but no one has had sufficient curiosity to study what causes it. Lecomte put about a hundred worker bees in a metal gauze cylinder and then put this on the floor of the big cage. A few bees were then released into the cage and gradually accumulated on the gauze cylinder. The proof was there: there is an attraction, but of what nature? Moreover does it need dead or live bees? Living, obviously, because freshly killed bees put into the cylinder do not interest other bees. What are we going to do now? This was what Lecomte was thinking that day as he held a test cylinder of very much alive bees in his hand, and they "vibrated". It cannot be said that they buzzed under these conditions: it was more a kind of sustained vibration, not audible, but nevertheless felt when you put the palm or the back of the hand on the little cage. Perhaps this vibration was of importance? Another failure: living bees enclosed in a hermetically sealed metal box did not attract their fellows. There was only one hypothesis left; smell *alone* cannot be effective since dead worker bees do not attract; neither can vibration *alone*. What could be said of the two factors working together, smell and vibration? Lecomte then put a handful of dead bees on top of the sealed metal box containing the live bees with complete success. Other bees introduced into the cage flew towards the "synthetic" bait made in this way.

This then is how they attract each other; but when you start digging into the problem it seems to be more complex, at least from the point of view of smell, as we shall see below. . . . It has been known for a long time that a bee that tries to enter a hive other than her own is very roughly handled; this means she has been recognized as a stranger, but how? As the guards feel the intruder with their antennae, it was thought to be a question of a characteristic odour, the hive odour peculiar to each colony. Reason obliges us here to postulate a considerable number of components of this smell, so that their mixture in varying proportions allows each colony to be distinguished from all others. Ribbands has advanced a very ingenious theory on this point. It has been known for some years that two hives placed side by side do not collect the same pollen and nectar, that is to say they do not visit the same plants, or do not visit them in the same proportion. This can be seen by means of a new method of microscopy now much perfected, pollen analysis. Everything bees collect in fact contains some pollen grains; honey always contains a certain number. A number of specialists, have issued pollen atlases. Palinology is used to identify fossil pollen, for nothing keeps so well as pollen. One can thus study the flora and seasons of the stone age with some exactitude.

When pollen traps (see below) are used to collect some of a hive's crop it is clearly seen that its floral make-up is not the same as that of the neighbouring hive. The differences are not small but, on the contrary, are considerable; they can be seen at the first glance, even if it is only by the very different colours of the grains. Thus, says Ribbands, it may well be that the skin of the bee, which absorbs odours very readily, gets impregnated with the particular floral odour corresponding to the composite odour of the flowers from which it is collecting. The infinite number of combinations possible thus forms a logical basis for the individuality of each hive's odour. A very interesting theory, so seductive in fact that one would almost swear it to be true.

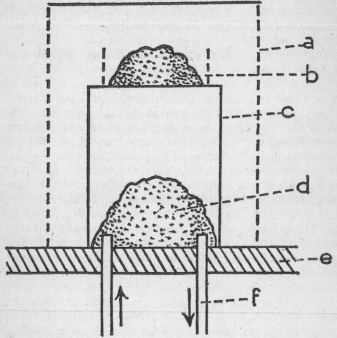

Fig. 10. Lecomte's experiment on attractivity among bees. Living workers are enclosed in d and supply the vibratory stimulus. The enclosure c is of metal and completely closed (it is aerated by f): dead bees are placed at b and supply the olfactory stimulus. e is the base and a the outside wire mesh. Workers from outside collect on a.

Science is full of theories so well-conceived and so satisfying to the mind that they appear to be stereo-isomers of the truth. Everyone then sticks to it up to the limits of reasonableness. How could one possibly believe that the planets moved in ellipses when the circle is the most perfect figure, the image even of divinity? And the scientists, whilst waiting for Kepler and Copernicus, tried to remove this shocking theory of planetary ellipses by the stupid one of epicycles. Our bees presented a similar problem, though on a more modest scale, for though we found ourselves fascinated as much as the others by the theory of the floral mixture, yet certain experiments made by Lecomte cast a shade of doubt on it and led us to look at the matter a little more closely. If you take a hundred or so bees from *any one hive* and put them into a cage on different sides of a sheet of glass dividing it, you can see that if on the first day you draw out the glass the bees mingle without difficulty; but if you do it on the second day fights start, which become fierce and end in extermination if the mixing of the original sisters is put off to the fourth day. Now the main point is this, the aggressiveness is *exactly the same whether the bees have been fed the same or different honey.* Now the theory demands that in the first case (fed the same honey) their skins would be scented the same. But the experiments show clearly enough that this does not stop a massacre. Let us stop there, for this is the central problem. How is it that two halves of a population of sisters can become strangers and enemies in a few days? Note, however, that all the bees making up the population are not necessarily sisters, but only half sisters. This is why: we used to believe that the queen went out but once in her lifetime and mated but once with a single male. But on the one hand the Americans and on the other the Austrian Ruttner and the Pole Woyke have proved quite clearly that the contrary is true. The queen goes out frequently and mates with five or six males. The spermatophores with which the queen's seminal vesicle is filled to bursting point do not mix but are discharged one by one: this is why in a population of black bees yellow ones can suddenly appear, the issue of a different male parent. Thus nothing is more heterozygous than a bee population in a single hive; it is pretty well as much so as a group of men picked up by chance in the street. But still, you might object, such a group of men, separated into two parts for a few days, would not fly at each others' throats when reunited. I agree, unless the separation was radical and lasted a long time. Then perhaps group traditions and habits might have been formed and hostility towards another group could have been produced. But what sort of group habits and traditions could the bees have shaped? Their life is much shorter than ours, since the foragers hardly go out for longer than three weeks or a month and die at the end of that time. Everything, then, goes much more quickly with them than with us; but I do not for all that believe that we must attribute the quarrels to changes in habits formed very quickly and markedly before the two groups were reunited. Lecomte and I thought of a much more practical explanation, one based on the everlasting exchanges of food so characteristic of social insects. Whether one feeds the two halves with the same or different food makes no difference: for it is not the food itself that is important, but rather its elaboration in the social organization, followed by the continual levelling of the products of the bees' metabolism by the above-mentioned constant buccal exchanges. And when two groups are not homogenous (and there are but slender chances of their being so among bees, as a result of differences of age and ancestry), their group metabolism is compelled to produce different odours that they cannot exchange (and thus equalize) with the other group. This is the explanation we propose: Lecomte maintains that it is still only a hypothesis, but it seems to be the only one that makes up for the deficiencies of Ribbands'.

There are still a few difficulties in the way of understanding the phenomenon of mutual recognition. For example, a bee which enters a strange hive is examined carefully by the suspicious guard bees, and at a certain point something subtle happens, which Lecomte suspects but is not yet able to define very clearly; something in her attitude changes and it immediately provokes the furious attack of the guard bees. It must be a question of movement, perhaps the creature is a little too "nervous". Another of Lecomte's experiments seems to prove it: he put a small group of bees into a cage and then moved "under their noses" the body of one of their companions. Whilst they took no notice when the body was immobile, when it moved they attacked it violently. Moreover if one marks bees by means of coloured dots it is soon seen that it is always the same bees that attack, as if they were "soldiers", although not recognizable from their morphology. In fact these attackers are the oldest bees and aggressiveness increases with age. Age is not the only internal determining factor, but the matter would take too long to go into here.

THE BEES' ROUTES

These very fine observations were made entirely in our laboratory. Lecomte's great merit was that he supplemented them by field observations. When one jiggles a little bit of paper or a twist of cotton wool in front of bees they angrily attack. By varying the bait one can find out what particularly annoys them. These are small things, about the size of a bee, dark, hairy and smelling of venom or sweat. But if these same experiments are done near a hive the attacks are almost always made in front of it, that is in front of the exit hole; this is why beekeepers always approach a hive from the back. But before getting to the hive or on leaving it the bees follow a predetermined course, one leading to the nectar-bearing plants already discovered by the scouts. These paths are about ten metres high and two or three metres broad and are so clearly defined that one can make a map of them. Now if you send a piece of brown wool, or the body of a bee, attached to a little balloon along one of these routes the foragers will furiously attack it. A metre away from the path they are no longer concerned, though they can easily see the offending object since it is being agitated; they will not deviate from their path to sting it. Consequently the degree of aggression shown serves to map out the bees' routes. The strangest thing is that they remain the same from year to year, because they are related to the topography of the terrain—for example, taken to a spot in front of a forest they will "aim" at a clearing, which enables them to pass lower down and thus avoid having to fly above the trees. They are not unaware of the law of least effort! As long as the topography of the neighbourhood is the same, so are the routes the bees follow. The fact recalls the paths trodden by ants, which have also been mapped and found to be very constant.

BUILDING BY BEES

The bees' constructions have always appealed to man's imagination. Nothing is more beautiful than a piece of virgin wax with its milk-white colour and geometric form. Réaumur proposed that the width of a honeycomb cell be chosen as the unit of measurement; but this unit would hardly have been accurate, because, apart from the fact that the size depends on the race of bees, there is a gradual increase in the size of cells, from that of the workers to that of the largest male cells. At about the same time a mathematician, Maraldi, was given the following problem: what would be the shape of a vessel giving the greatest capacity with the least use of material? And Maraldi replied: the solid hexagon.

Some years ago, the Abbé Darchen and I were thinking of this as we turned in our fingers this wonderful wax. He was looking for a subject for study; why not try construction in wax, I proposed. He accepted the idea.

It is always dangerous to propose a subject for a thesis. No doubt the experience a director accumulates over many years gives him a kind of flair; he has a confused feeling that he is proposing a good subject that will enable his pupil, in a few years, to get enough new and interesting results to earn him his doctorate. But chance can play a part; with this cursed biological material full of pitfalls and blind alleys one never knows. Suddenly one feels, in advance of the pupil, that the thesis is taking shape; but one may well have to wait a long time for this moment. What struck Darchen and me was the shape of the new comb; an ellipsoid with a waist near its original point of attachment to the wooden frame, with narrow and always absolutely regular sides. It is only the size that varies according to the number of bees working on the particular task. How do they set about it? We do not know, because everything is done at the centre of a close mass of bees, the wax-making groups, the centre of which reaches a temperature of 34°C or more. It is in these biological ovens that wax is made and built into comb. The bees there are strangely immobile, hanging by their feet, sometimes forming darker streaks corresponding to a thicker layer of insects. Some authors maintain that from time to time a worker leaves the bunch and goes and places her grain of wax on the comb (the wax having been secreted by her ventral glands and hastily formed by her mouth parts). After that she takes her place again in the wax-making group to receive there, perhaps, directions as to the construction to be made, whatever the form and mechanism of such direction may be. But it is either more complicated or simpler. For a long time we were looking for a method of reducing the thickness of this mass of wax-making bees that hides everything from us, without upsetting the process too much, and not long ago we found out how to do it; the nub of the problem was in fact the high temperature prevailing at the centre of the mass; perhaps the greater part of the workers were there for no other purpose than to raise the temperature and *if the bees were supplied with an exterior source of heat* the mass of wax-makers might be less dense. With this in mind I had a very flat box made with two glass sides, leaving scarcely room for the thickness of a comb between; this was put into another warmed box. Under these conditions the bees worked very well and quickly made a fine comb, although there was no thick wax-making swarm; it was very strange: the bees formed themselves into what I can only describe as "garlands" of varying width and we could foretell where the following day's building would be by following the contour of these garlands on the glass and marking it with a wax pencil; they traced out to some extent the next day's work. When the comb appeared the wreath of bees was always at a certain distance from it, or rather the thickest fringe of unmoving and tightly massed workers was connected with the comb only by a loose network of comparatively few bees, and the actual builder-bees did not necessarily come from the "garland". Darchen was able to assure himself of this by spending long hours in front of the observation hive and noting what the bees were doing, one by one; for example a worker, come from goodness knows where, encountered the wax group and seemed to pay but little attention to it, but nevertheless, marching over the bodies of her immobile sisters, she put down her farthing's worth of wax in the right place. After this she might join the mass of bees or go right off again. Do the garlands of bees, then, play the part of indicators, of punched cards that give the outlines of the work to be done but are nothing without the workers who use them? There is no doubt that these instructions concern the generation of extra heat for the zones where building is to be done, as Darchen's recent experiments seem to show.

That was an occasion when we were able directly to see what happened. In another experiment we were less lucky, nevertheless the mystery of the wax-making mass and the chains of bees was illuminated from another angle. It all arose from one of Darchen's strangest observations. It is known that combs are parallel to each other and we shall see below the hidden springs of instinct by means of which the bees maintain this parallelism. Now if one puts a little sheet of wax *perpendicularly* between two combs it will be found, after half an hour or an hour, turned round and made parallel to the comb. This is not just a fortuitous event; on the contrary, it happens every time. A strangely disturbing phenomenon, for it must be borne in mind that bees have never had any occasion to do anything like this in their hive, at least in this form; yet their building instinct at once seems to have found the right solution to the difficulty. Once again we went back to find out more of what was happening. This time we used a very flat hive, flat in the other direction if I can put it that way; it had a series of combs only a few centimetres high. It took us two years to get this conjuror's contraption ready to hold the bees: we had to heat it from outside, so that the wax-making masses would not be too thick. The ceiling and floor were of glass; a lamp placed a few centimetres below the floor enabled us to see the garland of bees. What a strange sight it was! With the typical ingenuousness of man we thought that one body of workers would have to fasten itself to and work from the sheet of wax on one side and that another body would work from the opposite side and that by pulling in opposite directions they could bring it parallel to the comb. But this is not at all what we saw: there appeared in silhouette, lit by the lamp below, a network or irregularly sized meshes, made up of chains of bees attached on one side to the sheet of wax and on the other to the comb. The meshes of the net gradually changed, following some law not yet understood; finally the sheet of wax reached a position parallel to the comb. Suddenly I found myself reminded of the nervous system and of the way its branching fibres form a network too; and above all of the most developed part, the brain: and in the brain the *reticular system*, the three-dimensional network through which everything passes and is either co-ordinated or cancelled. Then I thought of the network circuits of the electronic calculators. Networks seem to be an essential element in the working of information mechanisms.

But let us get back to the little wax ellipse that the bees start making when you put them in an empty box. It inevitably reminds one of a living tissue, of some support, such as a cuttlefish bone, which has the same shape. Now it is a law of life that tissues repair damage; Darchen, for instance, broke these foundations in various ways and always found that these "wounds" were "healed" very quickly. This gave him the idea of stopping the "healing" process. All he had to do was to put any small object, a match for example, in the base of the ellipse; this produced a bulge in the construction which only merged into the rest much later. A complete cessation of building could be obtained by putting in a narrow slip of metal a centimetre wide. Then the wax construction lost its symmetry and was only built on the opposite side, except when the metal was pierced with a certain number of holes the diameter of a bee's body. Then the workers cautiously put their heads through the orifice, then their bodies, building started again and the metal plate little by little became part of the comb. Darchen has even seen bees put their feet through the hole and place them against the feet of the workers on the other side.

These are strange phenomena. There are others no less strange which concern the distances between combs. In fact these are generally equidistant if the bees have enough unobstructed space. But what happens if a sheet of wax is put between two combs in such a way that it is too near one of them?

In such a case, if the critical distance has been overshot, the workers bring the wax sheet back to the right distance by pulling in the opposite direction. However they can only do this with a sheet marked with the honeycomb outline, that is a sheet of

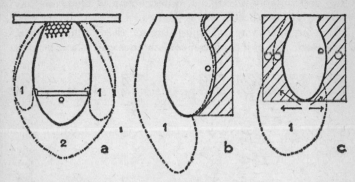

Fig. 11. Building among bees, after Darchen.
a An obstacle (o) has been placed longitudinally on the lower part of a comb: it leads to the production of two abnormal lobes (1) which are not united until later (2).
b A metal plate has been put on the right of a comb: it completely stops building, which continues only on the left (1).
c On the right a single hole in a metal plate is not enough to lead to building; two holes are needed (as on the left), near each other, before social co-ordination and building can start again.

foundation comb; a comb placed too near cannot be displaced in this way. In this last case the nearer cells may be reduced, but this is a much less usual action. Darchen noted that it was only a comb placed too near, and not one too far away, that led to such reconstruction. But what are the stimuli that induce this action? Another experiment can tell us that. If we put a sheet of wood or cardboard on the surface of a comb that is too near a foundation sheet then the bees do not try to remove the sheet of wax but hasten to bring it nearer and stick it to the screen. This only happens if the screen reaches the lower free edge of the comb being built from the sheet of wax. If it only hides the too-near comb up to the level of its base, stuck to the frame, then nothing happens. In a comb, as with a living being, there are some particularly sensitive, rapidly growing zones: such are the edges of comb, particularly the lower edges.

However, one of Darchen's most intriguing experiments was to *put some windows in his screen*. If these windows are of sufficient size and suitably placed the reconstruction of the comb may start again: the sheet of wax is then no longer stuck to the partition but built away and adjusted to the right distance. Then, you might ask, is it the presence of another wall of wax near by

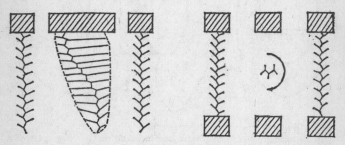

Fig. 12. *On the left.* The bees build a curve into a comb, which has been started with a wax foundation placed too near to a comb on the left; the comb is then built away to the right resulting in the depths of the cells being very abnormal. The workers try to re-establish the distance and parallelism of the cell openings as they proceed. *On the right.* A wax foundation put at right angles to the neighbouring comb is turned through 90° in order to restore parallelism in the combs.

that leads to this adjustment of distance? Not at all, for if you cover the partition with a layer of wax nothing is changed and the foundation sheet will nevertheless be stuck down. *The wall must be formed of cells* for the subtle distance mechanism to come into play.

It must be added that if a comb is deformed in such a way that an angle results, the bees will be able to realign the partitions and put them in another place so as to get the external face of the new cells parallel to the established comb. The cells then take on a strange shape; some are very deep and others are too short. The bees know how to remedy this defect by moving the base, but it is not often done and must be rather difficult.

If we now seek to find out just how the bees set about this very precise operation we run into a number of difficulties. In so far as we can distinguish what is going on, and this is pretty difficult, we only see our old friend the garland of worker bees hanging on by their feet, almost immobile, and joining one comb to the next. And that is all we see. Such chains play a yet stranger part in another experiment thought up one day by Darchen. Hives are usually vertical, as are combs. What would

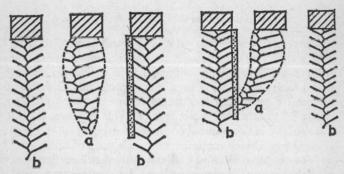

Fig. 13. Experiments in parallelism of comb. The three left-hand figures:
a Comb shifted to the right; cardboard sheet to the right.
b Re-establishment of parallelism.
In the three right-hand figures the cardboard screen is too near the foundation, which has shifted towards the left: in this case the bees fix the foundation to the screen and there is no adjustment of spacing. Wax made up into cells is essential for the adjustment of comb distance. (After Darchen)

happen if they were steeply inclined to one side? Nothing very much in the case of old comb with thick wax, but in contrast young comb, with soft wax, shows an annoying tendency to sag, at least at its lower end, and come to rest against the comb beneath it. However the workers are watching it. They try to support the sinking comb to keep it in the correct position, and they even put in wax supports for this purpose, attaching them to the comb above. We can cause still more trouble: if the spacing of the comb is maintained by means of a wire framework there is nothing to prevent one turning the hive completely to the horizontal. The comb is now in an entirely abnormal

Fig. 14. The regulation of building among bees. A frame has been bent into the shape of an inverted V. To restore the parallelism of the combs the bees lengthen some of the cells and drop the bottom of one cell at the top of the V.

position. Half the cells are pointing upwards and the other half downwards. This is a comb position the bees have never encountered in all their history, although it is familiar to another, South American, genus of bees, the *Melipona*; this genus, which also makes honey and has the advantage of having no sting, makes only horizontal comb, in fact. What happens, then, to the *Apis* hive in this unusual position? Nothing at all, strange as it may seem. Honey continues to be collected and brood raised with no trouble at all. And I tell the truth when I say that honey is stored not only in the upper layer, *but equally in the lower one*, where the cell openings point downwards! I had no idea how the droplets of fresh and very liquid nectar could be put into these last cells, until one day I tried putting a little water into a cell by means of a pipette and then turning the comb. If it is properly done, that is gently, nothing happens. The adhesion (surface tension) of the liquid is sufficient to keep it in place there. The bees' powers of adaptation are truly remarkable.

I will come back to this later, but first I want to tell you a bit more about the bees grasp of size and measurement. There is a really devilish way of testing this and once again it was the Abbé Darchen who found it. Let us suppose that we cut a comb vertically into two halves and that we put each of the halves into a sliding frame. Let us now bring the two halves together again but in such a way that the *space separating them is not quite right*, that is to say that perhaps the distance between them can be half a cell or a cell and a half. The bees make haste to join up the comb, with a multiplicity or irregularly shaped cells of all sizes. Next the *retouching phase* starts, which is very important and which we must go into in greater detail below. The problem admits no solution and the bees seem to realize it. This joining zone persists for a long time, with irregular cells, both larger and smaller than normal, which are destroyed and rebuilt several times by the bees.

They also carry out this retouching when a cell is carefully moved by an experimenter to a base which does not belong to it. Again it was Darchen who teased the bees in this way. If a hot knife blade is slipped along the base of some cells it is easy to remove them from their foundation and put them on to *another base, not belonging to them.* This is suddenly noticed by the bees; if, for example, some male cells have been put on to a worker comb the bees will laboriously try to reduce the size of the former cells in order to adjust them to the size of the foundation; of course this is impossible without starting again from the beginning. The workers seem to recognize this too, for they keep touching up the work; here and there we find abnormal cells, and even cavities in the wax with no opening to the outside, etc. In any case *it is the foundation that is important and controls everything,* and the workers are sensitive to the least disturbance. For example, I remember that I once made some experiments with solider and more robust comb. I soaked some strong paper in melted wax and formed it into comb on a machine. To my great astonishment it was a complete failure; the bees destroyed almost the entire comb thus made, as if they had noticed something strange about the resistance of the cell bottoms and sought to get rid of it. . . . The importance of the base explains the improvement brought about in apiculture by the manufacture of foundation comb. In practice, if a swarm is put into an empty box it has an annoying tendency to build mostly male brood comb and only a few worker cells. On the other hand the introduction of a little foundation comb, with the faint markings of ordinary brood comb, leads to the production of worker cells.

Is the influence of the foundation all-powerful, then? No, we can see it disregarded on at least two occasions: first, when the time arrives the bees will build male cells, even if the foundation is set for workers; secondly, if the large male cells only are introduced into a hive, it will soon be seen that the workers restrict them at the top, and shorten them too, so as to adapt them a little better to the needs of the workers they intend to raise there.

·UNUSUAL REPAIRS

Cells, then, can be modified in all possible ways. They can also be repaired or made good when unusual things have been put into them, a pin for example. If it has pierced the wall of some cells in a direction parallel to the bottom of the cells, the bees will cut the wall away until they reach the pin, draw it out and fill up the holes with wax. If the pin has been put right in the middle of a cell, perpendicular to the bottom, they will get it out by pulling it upwards, unless it has been bent to one side. What happens then? What would you do if one day you found a thick column you were incapable of moving in the middle of your dining-room? Well, the bees solve the problem in a curious way: they bend several wax walls inwards so that the pin is surrounded with a partition and does not occupy such a disturbing position in the middle of a cell.

Obviously the solution is but partial and provisional, for some cells are too wide and some are too narrow. So much so that the bees take up the problem (to which there is no solution) from time to time, breaking up one piece of comb here, making another there, in short trying to solve the problem by constant touching up. These remarkable repairs and touchings up are very characteristic of the bees' building methods: we must now go further into the matter.

REPAIRS

Darchen has made us rethink our rather too rigid theories about instinct: this is one of the most interesting results of his work. What now remains of the old idea of the infallible bees, mechanically building their cells to a pattern set once and for all, cells so stereotyped that they could be used as a unit of measurement? There is but the ghost of the idea left, because something much stranger has been put in its place: a "social body" (we shall see below the exact significance to be given to these words) which takes account of any difficulties and tries to find a solution to them, even when one is impossible; *which feels its way and touches up*: that is not at all the way a simple machine works, but an activity of a higher order (I dare not say an "intelligent" activity, first because things are much more complicated, as we shall see, and also because this adjective covers an abyss of ignorance and obscurity). The corrections made as building

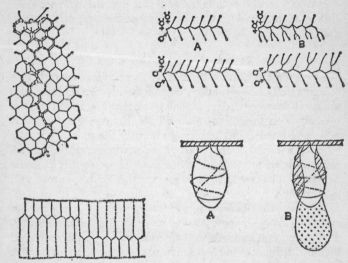

Fig. 15. *Top, left:* The repair of a longitudinal crack in a comb; solid lines, the first stage of the repairs; dotted lines, the second stage of the repairs.

Top, right:
A. Worker cells put on to bottomless male cells.
B. The results: immediately below A and B male cells are joined to the bottomless worker cells.

Bottom, left:
Variation in cell depth when one side of a comb is displaced in relation to the other. The dotted lines show the new work undertaken by the bees in order to re-establish the levels of the openings on both sides.

Bottom, right:
A thread has been wound twice round a comb at A: the area shown hatched is eventually removed by the bees; this allows the thread to fall off (after Darchen). The dotted area shows subsequent building.

proceeds can be noted from the foundation of a comb onwards; there are three stages: (1) the making of irregular, rough plans; (2) the carving of the whole into round cell-cups; (3) correction; one might almost say "camouflage", because the comb-foundations show very considerable irregularity (Darchen). We have already noted other examples of such touching up in the joining of two separated halves of a comb, in the way bees deal with a needle planted plumb into the middle of a cell, etc.

It is logical to conclude from all this that the bee should be able to adapt herself to all sorts of circumstances; and, in fact, she can.

ADAPTATION AMONG BEES

We have already noted a sufficiently surprising adaptation to a comparatively new situation, that of the hive on its side, with the comb horizontal. If a similar thing happens in nature, the bees abandon their home after a little, because the comb gets weak and breaks, and honey runs all over the place. It is only by using a few laboratory tricks, such as wire supports, that we can get the wax to stay put. Then the bees soon get used to the new position, as we have seen, and use the comb for brood and food storage.

Moreover they have no fixed ideas as to the proper places to store food. I do not know what whim it was that led me one day, during the great honey flow, to insert into a hive a piece of wood bored with a few round holes, irregularly spaced, about the size of a cell, at a time when nectar had already filled all the available cells. The only concession I had made to the bees was to soak it all in a bath of hot wax. And what do you think they did? Well, they just filled the holes in the wood with honey, because when the nectar flow is abundant, every possible receptacle must be used. This will only happen during a heavy flow. But nevertheless, some ten years ago, comb made entirely of thin aluminium sheet was on the market, having cells of the same size and shape as the bees' natural cells. There is no doubt that honey was quite happily stored in this metal comb and the idea was only dropped on account of cost and some mechanical difficulties. Recently the Germans have succeeded in moulding a complete comb in plastic and the bees not only put their honey there, but also raise their brood in this new material.

The most interesting experiments in the field of adaptability are those made by Vuillaume on the subject of the royal cells. We shall see presently that beekeepers induce the bees to raise queens in a queenless hive by putting in a line of cell-cups in each of which they have put a very young larva. The workers completely accept these cups, trim them to the "regulation" size and load them with an abundance of royal jelly. The cups need not even be in wax, those made of glass are accepted just as well and are topped with wax; the larvae get on admirably in them. The workers may also be offered plastic cups of various shapes and all goes well, with one exception—they do not like a square shape; the ends must be rounded. But there are many more strange things to be told about queen cells and Vuillaume's work on it takes us into some unexpected areas.

THE ROYAL CELLS AND THE EPAGYNES

Everyone has heard of the royal jelly madness of a few years ago, which so shook the bee world—in fact, quite simply, the world. It is more easily understood if a few figures are mentioned: royal jelly was sold at a higher price than gold, for it started at 1,800 francs (old) the gram. However, at the time it was 500 francs the gram more than 800 kilos were sold in France in a single year. It is thus easy to understand the enormous interest beekeepers took in it. It is true, moreover, that the substance has a certain therapeutic interest (for instance it contains a new acid, deconic, found only there). But it is unfortunate that an absurd publicity campaign spoilt the product in the eyes of the medical world. It should be noted in passing that it is neither as rare nor as expensive as it was said to be, for a single hive can yield up to 500 grams per year, using a new and perfected production technique.

This technique is an interesting one; I have already mentioned the subject briefly above. Little cell-cups of wax or glass (or indeed any other material would probably serve) are fixed to laths at the lower edges (an essential detail). In each is put a drop of diluted royal jelly and a very young larva taken from a comb. These laths of 40 to 60 cells are introduced into a queen-less hive full of bees, who will not hesitate to fill each cup with from 100 to 400 milligrams of a whitish substance resembling yoghourt, but with a particularly horrible, burning and acid taste. This is the famous royal jelly.

There are many factors to examine: first of all the form, size, placing and material of the cups, all things which can easily be varied. Vuillaume was not slow in this respect and one thing he soon found out was the famous interchangeability of the cell wall material. He also found that bees recognize the larvae of their own species: for instance they will reject ant larvae that have maliciously been introduced into these receptacles. How-ever they do not recognize the sex of their own larvae, so that male larvae are accepted and nourished in the cups. As a matter of fact such a thing may be found under natural conditions, where males can be found completely developed, and ready to emerge, in queen cells. The strangest part is that they manage to live under conditions not designed expressly for them; such males, in addition to being a little bigger than normal, also have a number of abnormalities in the genital system.

All this is certainly strange; it is one of those cases where nature gives us a hint and enables us to take advantage of a passing opportunity. At various times I have used the words "accept" and "acceptance". It is now time to define them a little better. The wax cell-cups introduced into the queenless hive are not all transformed into royal cells; a certain number are left alone. However the percentage abandoned varies accor-ding to the origin of the wax, and also whether such cups have previously been introduced into the hive or not.

The problem of the wax troubled us considerably. From the outside the blocks of wax from which Vuillaume made his cups all looked alike; some were perhaps a little darker than others, perhaps the pleasant smell they gave off was not always the same (this is wrongly called the smell of wax when in fact it is the smell of propolis). But these differences seemed very slight. And the worst thing of all was that the bees took to the glass cell-cups better than to the wax ones. Here was the hint! Here was a small clue that meant we must stop short to see where we were.

. A professional man, who had spent twenty years in a laboratory, could not fail to jump to the conclusion. Glass is a neutral substance, and it is most unlikely that it can influence the bees in any way. Hence if some waxes are less acceptable than others it is because *they contain a repellent substance*, in varying amounts, according to the origin of the waxes. There is no other possible conclusion. We now had the clue, all we had to do was to work along the trail, a chemical trail of course! I won't bore you with details; however you should know that wax can be extracted with several solvents, acetone for example, which, when hot, will extract some substances, and benzene, which will not extract quite so many. Now waxes extracted with acetone give cups that are almost entirely rejected whilst those made from the benzene-extracted wax are accepted without the least hesitation.

This means something to a biologist in any way familiar with the chemistry of extraction; it shows that we are dealing with a substance which scarcely dissolves in benzene but dissolves very well in acetone and that it is likely to be propolis. Could it be, then, that the variable quantities of propolis, always found mixed with wax, was the cause of the differences in acceptability of the waxes? It took no longer than a quarter of an hour, after the idea had occurred to us, to make a propolis extract and to sprinkle it over the royal cell-cups: but we had to wait until next day for the results. A day had never seemed so long: but the experiment was a hundred per cent successful. None of the cell-cups sprinkled with propolis had been accepted by the nurse bees, while the control cups, alongside, were brimming with royal jelly. How happy we were!

This seems to be the moment to deal more fully with the subject of propolis. The word comes from the Greek and means "in front of the city", and it designates a blackish, greenish or reddish gum which makes the fingers sticky. The very pleasant smell is a little like that of incense (and it is even used in the East to adulterate this product). The bees put it on the tops and side faces of the frames or of the hive. It is always mixed with a little wax; and it is from the propolis that the hive odour comes. The origin of propolis is pretty mysterious; bees can be seen bringing it back to the hive in the form of small sticky balls, attached to the hind legs, like pollen. But where have they got it from? From tree buds, there is no doubt, especially from the buds of poplars and the various willows. But this is where difficulties start. Because if you extract tree buds with acetone, in only one case, that of the poplar, will you get a substance having the smell and certain characteristics of propolis. For instance poplar-bud extract leads to the rejection of royal cell-cups as does propolis. This has led some authors to maintain that propolis comes only from poplar buds; but then what do bees do in areas where this tree is absent or rare? For propolis is found more or less in every hive and in every country; and consequently there is still some doubt as to its origin.

Well then, too much propolis in a wax makes it undesirable for the building of queen cells. But nevertheless even a good wax used for making these royal cells is not automatically accepted. Vuillaume noted at the very beginning that the acceptance rate on the first day of introduction is poor; if the lath is drawn out and the cell-cups filled again with young larvae the acceptance percentage rises on the following days. Vuillaume obviously wanted to know what would happen if the empty cups were left for a day or two in the hive *before being filled with larvae*. The results of the tests he made surpassed his wildest hopes: this "familiarization" of the cell-cups had the effect of multiplying the percentage of acceptances by three or four, always provided that the cells were not forgotten and left three or four days in the open air: in this last case the advantage possessed by the "familiarized" cell-cups lessened and finally they became no more acceptable than the new cups themselves. There must, then, be an "acceptance substance", some subtle odour no doubt, which impregnates the wax and renders it acceptable to the bees; however, this very important characteristic gained by the cups is evanescent and does not last long in the open air. *This acceptance substance*, or *epagyne* (see below), gave us a lot of trouble. Up to the present we have not been able properly to isolate it.

THE EPAGYNES

I have had many occasions to mention these odiferous substances: they appear to play a big part in bee behaviour. This
did not strike me until these last few years when I had occasion
to write a report on the work done in our laboratory over the
previous thirteen years. We had all given of our best and the
results were wonderfully productive and varied; so much so that
I felt a bit swamped by them. I could not see the wood for the
trees, which is bad when one is directing a research institute.
But, as usually happens, things became clearer as soon as I got
down to it pen in hand. How had I not seen this before? It was
obvious that here and there, both inside and outside the hive,
the bees put down a whole series of odour spots, either attractive
or repellent, which guide the whole colony in its daily life. It
was written as clear as daylight in the body of our publications
and we hadn't been able to see it. There was a complete odour
language which we were only beginning to learn.

So I coined a name for this group of substances. This often
happens in the natural sciences and is one of the reasons why
our jargon seems incomprehensible to the layman. But what
else can one do? It is best to make up a pleasant-sounding word
by running through the Greek dictionary. I made "epagynes"
from the verb ἐπάγω, "I put upon", because all these substances
give special qualities to the objects on which the bees rest.
When it came to making a list of all these materials, giving each
one a different Greek letter, I began to doubt whether the
alphabet would suffice.

There are some substances that the bees put down inside the
hive, as can easily be seen by introducing a piece of wood into
an observation hive. Such bees as first come into contact with
the wood start quickly away from it. Little by little this avoid-
ance reaction becomes less and finally it disappears completely.
At the same time the piece of wood turns colour and becomes
yellowish, and its smell changes too; it "smells of the hive".
What is the nature of the material that impregnates the wood?
We know nothing of it: propolis perhaps? In any case the above
forcibly reminds one of Lecomte's experiments on aggressive-

ness. When a stock of bees is put into a new cage, at first they show no hostility to a stranger worker put among them; for the hostile reaction to take place they must have passed a certain time in their new abode, two or three days. But if this cage is not strictly new, that is if it has been left *empty* in the hive for a few days before the experiment was started, then all will be different and a stranger introduced to the stock will be belaboured from the start.

There is also the famous familiarization substance for the queen cell-cups that we have just been considering; this, on account of its great instability, seems different from the first group of materials.

When the bees feed on a saucer of syrup, it is at once seen that the edges are rapidly soiled with a yellowish material. This is not excrement, because bees only defecate in flight, and moreover this material does not have the unpleasant smell of bee faeces, but smells more like wax. But certainly, if it is a wax, it does not come from the abdominal wax glands but from somewhere else on the body, perhaps from a special gland hidden at the end of the feet. In any case it possesses the property of attracting the bees from the home hive and repelling those of other hives. I could give details of half a dozen more substances if I were not afraid of making this section of my book inordinately long.

Let us turn rather to the "repellents", the equivalent to the bees of a "no entry" sign to us. One of them may well be propolis itself (or, to be more exact, a substance contained in propolis) which leads the bees to reject a cell-cup as unfit for a queen cell. Another still more curious substance comes from a wounded bee's body; it can be classed in the group of "fright substances", studied by von Frisch in connection with minnows. When these fish are attacked and bitten by a predator, their fellows fly from the wounded fish with all the speed their fins can muster, which is useful in that it gets them away from their enemy. The "fright substance" (*Schrecksubstanz*) is found in the muscles of the fish, from whence von Frisch was able to extract it. A few drops thrown into the water bring about a frantic flight of minnows. Now in the case of bees, if you wash the bodies of workers with alcohol, taking good care not to wound them, you will get a group of highly attractive substances (epagynes) of the same kind as those which distinguish the colony's nurse bees; but if the bees have been a little crushed before the washing with alcohol, then another group of substances is obtained, which causes a flight of the worker bees. I think the analogy with von Frisch's minnows is very close.

The same thing can be noted outside the hive; for some little time now it has been known that if bees collect honey, pollen and propolis, it is because these contain substances which attract them (they are called "allectines", from the Latin verb *allicere*, to allure, entice, attract). Pollen in particular has been studied and attempts at extraction have been made by Louveaux. Ether takes out certain substances that can then be mixed with any sort of powdered material, for example flour. In such a case the bees start to collect great balls of flour on their back legs.

THE PROBLEM OF POLLEN

The collection of pollen itself poses a number of problems, and Louveaux made this his subject. It is a fascinating one, but to understand it one must have seen the pollen traps in action on a fine spring day. Their construction is simple: it is just a question of a metal or plastic sheet bored with round holes about 4 mm. in diameter. The bees easily get through them, but the big balls of pollen on their back legs are caught by the edges of the holes and fall into a trough beneath. In this way one can collect unbelievable quantities of pollen, more than 100 grams per hive per day. The pollen spheres are of every colour, from white to sky blue, and run through green, yellow, violet and black. The bees collect it from a certain number of flowers which they assiduously visit; there are no two hives, even those situated side by side, which collect the same coloured balls or collect them in the same proportions. Louveaux devoted himself for many years to the laborious collection of statistics on this matter. We used to rag him when we saw him endlessly labelling his pollen balls. But his patience was infinite, a thing so necessary to a scientist, and his devotion to the matter bore fruit.

In short he found that not only did two hives, even though side by side, collect very different pollens, but that also they collected much the same year after year; and moreover, groups of hives from outside the area, brought to the Parisian region, showed an individual behaviour, very different to that of the region's own hives: that is to say their choice of pollen-bearing plants clearly distinguished them from the Paris bees. Thus a hive devoted to willow in 1961 continues devoted to it in 1962 and so on. In itself this is rather strange, because it must be remembered that the worker bees have only a very short life, a month at most, in summer; the bees that see the willows flower in 1961 have been dead for a long time when they flower again in 1962. Only one insect in the hive has a life long enough to remember the passing years: the queen, who can live six to ten years; but we must remember that she does not leave the hive after her nuptial flight, and does not feed directly on pollen or nectar. The workers feed her from the secretions of their nurse glands. As she never goes out how does she know about the sequence of flowerings? Where is her "memory organ"? More-over does such a thing exist?

To solve this problem Louveaux had to follow a long and tortuous experimental road. He says: "First of all there is no real memory. A group of hives, coming from Provence where box is abundant and much frequented by bees, was designated by the letter P. This group was brought to Paris, where it should have shown a preference for plants such as *Buxus sempervirens*, much commoner in the Mediterranean region than in the north of France. In the event the P group showed very little interest in *Buxus*, but on the contrary took to plants such as colza, sainfoin, mustard, and red clover, all much less common in the Midi than in our region. *Acer pseudo-plantanus* (the syca-more), a mountain plant, was not particularly visited by the bees in group J (coming from the mountainous Jura). The hives in group R, coming from the Indre region, where sainfoin is abundant, were not interested in the sainfoin at Bures-sur-Yvette" (Louveaux). It seemed that some other behavioural or ecological factor, or whatever you care to call it, was at work. For instance, one might suppose that, according to their origin, a stock of bees might specialize in visiting the tops of trees, another the bushes, a third the low-growing flowers. Thus three different flight levels, stamped into the hereditary pattern of the race, might give rise to three different patterns of pollen crop. But in actual fact we did not find that any particular group of hives was always the one most interested in, say, bushes or herbaceous flowers, or flowers with some common ecological characteristic.

At this point another hypothesis was proposed, an unexpected one, arising from certain side issues concerning the nitrogen content of pollen. From Louveaux's many experiments the conclusion can, in fact, be drawn that the nitrogen level, and consequently the feeding value, of various pollens is very different. Now if the pollens are arranged in groups, in decreasing levels of nitrogen content, *it can clearly be seen that certain groups of hives prefer high nitrogen pollens and other groups of hives prefer low nitrogen pollens.* The persistence in collecting willow pollen, for example, that certain colonies show from one year to the next, is in no way a memory of willow as such but is a wish to collect the richest pollen: and in this group there is not much choice. In the same way groups of hives from other regions, brought to the Parisian region, *preferred for any given season to collect pollen with a certain nitrogen content,* and this distinguished them from the local hives.

It can be objected that this is only a way of shirking the problem. Who, in fact, feels this need of a certain pollen, the queen or the workers? Louveaux found himself in a position to answer this question by quoting the considerable differences in hive development. Colonies which like rich pollen and only visit a small number of plants are, in fact, the most vigorous and the ones to develop first, and vice versa. *So that what really varies between hive and hive and is implanted in the hereditary material is, in fact, a certain rhythm of development.* And as everything depends in the last analysis on the queen's rhythm of egg-laying, the key to the matter must rest in her body. Now what would happen if a Provençal queen were put into a Parisian colony, and the reverse? It is a question we cannot yet answer. The experiment is very difficult to do and as yet we have had no success with it.

ATTEMPT AT A CONCLUSION

Let the reader be under no illusions: the above pages may perhaps have seemed to him to be too full of detail and all too capable of exhausting the subject! In fact it is no more than a summary, a quick and very incomplete one, of some of the salient points in bee behaviour. Nothing much has been said, for example, of nutrition, nor of other matters, such as the physiology of reproduction, where progress in our knowledge over the last few years has been remarkable. But we might here and now attempt a summing up, a synthesis, of what we have learnt. What a strange picture will take shape before us!

Our starting-point could be a quick comparison of a bee society with a mammalian or bird one; but having reached the point where we are at present we can easily see that we are dealing with a very different thing in the case of bees.

In my opinion, and there are many biologists who think as I do, we have in fact studied a *new type of organism*. And by that I do not mean bees considered alone, one by one; then they are but ordinary insects, Hymenoptera, a group which also includes wasps and ants, but these social insects are distinguished by one essential characteristic: *not only can they not live when isolated, but more curious still, their death occurs a few hours after being placed in such conditions*. This is the case not only with bees, but also with ants and termites.

This fact, so strange and so badly explained up to the present —isolation leading to death—has not received enough attenttion: it is characteristic of social insects and of them alone. Does this not remind you of organisms with cells which cannot live for long once they are parted from the organism? At the other end of the ladder of life may be found the sponges, which show a similar characteristic, a yet more curious one. The body of a sponge is formed of cells having a fringe and a flagellum, which group themselves round a central cavity. Now one can force the whole of a sponge through a sieve, so that all the cells are separated and form a reddish paste at the bottom of the basin. Leave it alone: after a while, the paste starts to regroup itself into a more or less spherical mass with a central cavity; there you have the sponge again. Does that not remind you of Lecomte's bees who grouped themselves around a little metal cage containing some of their companions? Or again, of that wax-making group that can be seen so well in the observation hive; that almost immobile stock with dark streaks where the bees are thicker, which slowly changes as the busy workers enter and leave? Give the sponge cells a little more independence and you have something very like the bees.

Let us accept, then, this hypothesis of the hive as a super-organism where the individual bee is just a detachable part with little importance, almost without separate existence. Those wooden boxes at the bottom of our garden from which the foragers leave then take on a more disturbing aspect. Inside each one is fairly bulky animal weighing some 4 to 5 kilos (there are 10,000 bees to the kilo), provided with a supporting apparatus (the wax comb), and hermaphrodite reproductive organs: the ovaries of the queen and the testicles of the male; admittedly it is a matter of seasonal hermaphroditism, for the males only appear during the summer, to disappear later; but that does not matter, for we know of more than one case of other animals showing seasonal hermaphroditism. Respiration of this animal is assured by the fanning of the ventilating bees, who blow out spent, or too humid air, at times with sufficient force to flutter the flame of a candle. There is a circulation, in fact a very active one, as has been shown by experiments with radio-isotopes. It is true that it does not work through veins and arteries, but the *buccal* exchange of food and social hormones is a perfect substitute. The *production of heat* is one of the most important social functions, and it is characteristic of bees. In the solitary state these insects have a temperature almost as variable as that of other insects, but as soon as some thirty or so are put into a cage and allowed to get into a bunch everything changes. A fine-stemmed thermometer put into the middle of the stock soon runs up to 30°C and over. In a hive, under normal conditions the temperature is 34°C and it is kept within half a degree of that, as with man. Is it too hot? The bees wet the surface of the comb, ventilate it and leave the hive in large numbers. Is the temperature too low? This is no problem: at least as long as the workers have a stock of honey, for this is their only fuel (moreover no other substance is known, apart from glucose, which gives such an intense thermogenesis).

There remains then only one big and inevitable question if

we wish to push this suggestion of a super-organism to its logical conclusion: where is the nervous system? Where is the brain? Note first of all that all the bee's reactions are governed by the society. This is easily seen when we put bees to the *preferendum* test, so common in entomological laboratories. *Then the social body dominates the individual reactions.*

· Insects are very sensitive to temperature, humidity and light; all three directly influence their activities and development. This can be demonstrated by, for example, placing some insects on a long metal plate, cooled at one end and warmed at the other. Along the "heat gradient" thus formed the insects will place themselves in particular characteristic zones according to species: these areas are their *preferenda*. The same thing can be done for humidity, by putting a plaster bed into drying conditions at one end and wetting ones at the other; or with light, by means of a darkened glass lit from above which gives every gradation between full light and dark. Under such conditions all goes very well with most insects and you cannot imagine the number of papers such simple apparatus, so childish in appearance, has given rise to.

But use the bee and everything goes wrong. First of all, if you introduce a single one into the apparatus she shows nothing more than considerable agitation and will not place herself in any fixed thermic, hygrometric or light zone. Rather better results can be got with little groups of 30 to 40 workers, although very different from those obtained with the solitary insects. Thus in the thermic preferendum test the little stock will assemble at any spot of the gradient and as soon as their bodies touch the temperature starts to rise to 30°C; in the hygropreferendum apparatus the group comes to rest around 40–50 per cent relative humidity and in the photopreferendum apparatus no reaction to light is shown. But what happens under more or less normal conditions in the hive? There neither heat, nor humidity, nor light seems to affect them. Full sunlight can bear down on a stock without their taking any notice, always provided that it does not get *too* hot; Verron placed a series of wet sponges above one half of a hive and a number of broken bits of quick-lime around the other half, which last considerably lowered the humidity on one side of the hive, but nothing worth remarking upon was seen and the workers passed indifferently from one zone to the other.

However, the stormy influence of the group on individual reactions tells us nothing about the "social nervous system", even admitting that such a thing exists. A recent study by the Englishman Vowles has drawn attention to the smallness of the nerve centres in this insect and the small number of cells they have compared to the large mammalian centres: it seemed to him that this should limit their psychic potential. It is certain, contrary to what is still commonly believed, that psychism is not in exact and direct proportion with the weight of the brain, either in man or in animals. But all the same, on the whole there is some sort of rough correlation: an ant, which has far fewer nerve cells than a rat, cannot be so flexible in its behaviour as that creature. *Nevertheless there is an exception: the social insects.* If, in fact, the little brains can *interconnect*, pool their resources and all work together they can then work on a far superior level. All the more in that there are over 60 to 70,000 workers in the hive, and as many brains as there are workers. To be better understood I must make a comparison with the big electronic calculating machines. We know that their memories are made of rings of ferrite joined in a complicated system. Let us suppose that an engineer, told to make one of these machines, had but one ring of ferrite: he could do nothing. If he had ten or a hundred he would be hardly any further advanced; but if he had several thousand, then he could connect them up and make the machine's memory out of them. A thousand units acquire a value and significance of their own which ten or one hundred do not have. Let us now suppose that the little ferrite rings are provided with legs, move about and only come back to join the whole on special occasions: you would then have a machine very like a hive.

Naturally such an argument must not be pushed too far; but nevertheless it has a kind of internal logic that makes it plausible. It is certainly true that up to the present comparisons between machines and living organisms have been defective because our machines have not been sufficiently elaborate, being based on insufficiently refined principles and nowhere approaching the complexity of living organisms. It is absurd to compare an animal with a locomotive or a motor car; but a predator pursuing its prey perhaps begins to resemble, though but crudely, a modern rocket fitted with a homing head. So, as I have just shown, do bees resemble a computer.

CHAPTER TWO

WASPS, ANTS AND TERMITES

PASTORAL AND WARRIOR COMMUNITIES

ANTS

SIX THOUSAND species of ants, all social, are spread over the surface of the globe. They are, without doubt, the only insects that can successfully stand up to man. The Americans are at their wits' end in the fight against the "fire ant", *Solenopsis saevissima*, which devastates their crops; in the West Indies and South America it is hard to prevent the redoubtable *Atta* from cutting all the leaves in an orchard to supply its fungus gardens. And even in France the Argentine ant has invaded the Mediterranean coast, from which, moreover, it is in the process of driving out the other species of ants. It is to be found everywhere, in one's bed, in jam-pots, of course, and in all stored food; not that it is dangerous or that its sting is unduly painful; it is quite simply exasperating. And how far can we defend ourselves against it? The colonies, which are numerous, live in the thickness of the walls and contain a great many queens, each perfectly capable of re-forming the colony. It seems practically impossible to get rid of them. Poisoning by sweetened toxic solutions has been tried. Alas! the bees hastened in multitudes to lap it up, and died of it. The next step was to keep the bees away by covering the bait with a metal sheet perforated with holes only just large enough to admit an ant. The ants perished sure enough, but nothing had been gained. As with bees, each group of ants has its strictly defined foraging area, and those whose area included the poison came to grief. The innumerable other groups foraged elsewhere and lived on. It is impossible in practice to get rid of them all.

In fact it seems that we must resign ourselves to living side by side with the Argentine ant even while we curse it. Luckily, it appears willing to confine itself to the Riviera for the time being, but a few years ago it was a near thing. A colony managed, no one quite knows how, to penetrate the Pasteur Institute; the scouts quickly spotted certain cultures which seemed to have been left there for the special benefit of ants and they lost no time in perforating the corks. What was the horror and amazement of the Pasteurians one fine day when they discovered long columns of ants boldly emptying a flask and trampling the frightful germs cultivated there all over the Institute—without the worker-ants giving so much as a shudder!

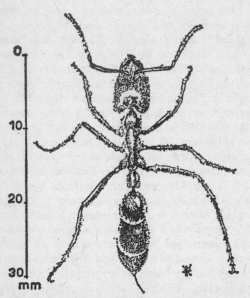

Fig. 16. Shown side by side are the largest and the smallest ant species in the world: the gigantic *Diponera* and, on the right, the tiny *Pheidole*. Both come from Iquazu in northern Argentina. (After Goetsch)

BUILDERS

In spite of everything, I cannot help liking them. I have always been more myrmecologist than apiologist, owing, no doubt, to countless afternoons spent in woods, watching columns of red ants. It is very hot; there are no human sounds, only a sharp buzzing that must come from bees foraging in the tree-tops; at ground level, by listening intently, one can just make out a soft little crepitation, a "rippling of water on moss"; it is the song of the tiny black wood cricket, *Nemobius*, running among the dead leaves in their thousands. Then there is a scraping noise as regular as if a battalion of pygmies were on the march: here come the red ants along one of their trails, probably on their way back from licking up the sugary excrement of aphids on the big pine tree. They pay no attention to me unless I approach to within a few centimetres of them, when a section of the column will halt and the workers raise themselves on their four hinder legs, the thorax vertical, the antennae vibrating. If I move, the abdomen curves up and aims a jet of formic acid at my face. If it touches the eyes it feels as if a red hot poker had been plunged into them. However, there is no need for all this anxiety: we can keep our distance. But what boundless activity, and how strange it is! What am I doing here, in the midst of all this heedless Nature? Ten metres away there is the ants' nest, a high mound of twigs. I have drawn a map of the many trails converging on it. They cover more than 700 metres. The sound of innumerable ant feet and the unique ant smell are perceptible at a distance of several metres. It is as if it were a civilization, very distant from our own, fallen perhaps from another planet, and with which there is no possible means of communication. No layman that I have invited to observe ants with me has ever failed to pose this question: is it you who are observing them, or they who are observing you? "All the same," I answer, "they lack one of the most obvious signs of intelligence: the power to progress. For as long as ants have existed (more than forty million years if Science remembers rightly) they do not seem to have advanced at all and the scanty fossil evidence indicates that their activities have always been much as we see them today." This reply generally reduces my inquisitor to silence—and just at the moment when I am beginning to ask myself whether my answer was quite honest and complete. For (a) one cannot really tell a great deal about ant activity from fossilized remains; (b) they are at least 40 times older than man as a race, since our stock hardly goes back a million years (the family of *Homo sapiens* properly so-called is much younger, about a hundred and fifty thousand years). Therefore their evolution, if it exists, must be much slower. Indeed, this appears to be true of all insects, since there is a cockroach of Permo-carboniferous times (350 million years ago) whose fossil remains bear a striking resemblance to the modern cockroach. Thus, if the two evolutionary speeds differ by a factor of 40, for instance, the habits of the ant might well evolve without us being aware of it. Or, to put it another way, it may be that 100 million years hence some learned scholar with six legs and two antennae will be writing that, according to very ancient tradition, it would seem that a certain race of monkeys changed very suddenly into another kind of creature that built goodness knows what . . . numerous mountains of stone . . . then disappeared like a flash of lightning. . . . However, in science it is useless to set oneself questions that cannot be answered.

But let us not be so severe with ourselves! One can dream. For instance, we can ask ourselves what the true meaning of polycalism is. This rather pedantic-sounding word describes a fairly well-known phenomenon in the ant world which is of special importance among the wood ants: it is the union of several colonies, separate yet maintaining a mutual relationship. In multi-queen species daughter colonies endowed with several queens, are formed at a certain distance from the mother nest (some tens of centimetres) and prosper like so many branches issuing from one main stem. With the red wood ant, however, this phenomenon assumes special importance. Raignier has described communes containing *several hundred* nests, stretching over ten hectares and connected by very constant trails which have been mapped. These colonies "know each other": that is to say that, inflexible and ferocious as they are where strangers—even of their own species—who enter the nest are concerned, these colonies will accept any member of their own polycalic community. Sometimes the nests of one particular section will be abandoned, but only for ten others to be built a little further on, as if the ground were being utilized methodically according to its potential value. You may say:

how is that any more mysterious than the polycalism of nests separated by a few centimetres that you mentioned just now? I agree: distance is not very important, but here the phenomenon is, as it were, magnified and made more striking. I also believe, as I have already said of the bee, that when the tiny brains of insects succeed in interconnecting, something quite new emerges, which is the insect community. There are several good reasons for regarding this as a super-organism. *What happens, then, when these multiple entities federate among themselves?* Will any engineer contradict me when I suggest that the capacities of a calculating machine depend on its size or, at least, on the number of elements it contains? (I enjoy these reveries: they are not as crazy as they seem; they nourish science which, in its higher manifestations, *blends with the spirit of adventure.* I say they nourish it, because they inspire new experiments, which is indeed the principal merit of a theory and perhaps its only one.) For example, are the links between the different components of a multiple society isotropic, that is identical in every part and in every way, going and coming? Or is there a hierarchy; does one colony predominate over the rest in one way or another?

We could find out, perhaps, by studying the exchanges, if there are any, along the well-worn trails that constitute, as it were, the commune's circulation; by the use of radioactive tracers, for example, served up—literally—in every sort of way, employing physics, of course, but also biology. Not only are isotopes administered to an animal or even a man, in order to trace their subsequent passage with the aid of a Geiger counter, they are also fed to insects, such as flies and mosquitoes, in an attempt to learn where they go after they are hatched and the extent of their dispersion. Furthermore, insect larvae have been dipped in radio-phosphorus and offered by the plateful to carnivorous, subterranean ants in order to discover the layout of their galleries, a ruse that has worked perfectly.

Unfortunately there were no polycalic colonies at our disposal. All the same we decided to study the comings and goings of an ant nest by this means, if only to get our hands in. So one fine day myself, Lecomte and a few atom scientists, with very impressive equipment, drove out to a wood near Epernon where I had been in the habit of teasing the ants for several years past. I chose a fine nest of the *Formica polyctena*, lying on a hillside and surrounded on all sides by numerous nests of the *Formica rufa*, a closely related species but monogynous (with only one queen), while *polyctena* is polygynous (there are sometimes 2,000 or more queens in each colony). This is the big ant-nest of which I spoke just now; the activity there is almost terrifying. It was destined to give us a number of very interesting results. The first, as biologists, we awaited with mischievous anticipation: we looked forward to observing the deportment of the two young physicists when they became aware that the *polyctena* were insidiously invading their trousers. In these conditions a biologist may preserve a shadow of sang-froid, but not so our votaries of the cyclotron. When they had adequately consigned us and the ants to all the devils in hell, the experiment began. First we poured into a saucer enough radio-phosphorus mixed with sugar syrup to assassinate three or four men; but the ants were not in the least concerned; they are all quite resistant to isotopes and will doubtless survive, untroubled and alone, an atomic war.

Up till then nothing out of the ordinary had occurred. The measurements along the ant trails had shown us, as was to be expected, an "activity" (to use isotope jargon) spreading like an oil stain at a speed relative to the amount of traffic on the different paths—which varies very much. I do not know which of us it was that suggested verifying as a control the "activity" of the neighbouring *Formica rufa* nests. Nothing was abnormal: that is to say, the needle of the Geiger counter registered only background noise, the invisible shower of cosmic rays that penetrates us all and the "emanation" rising from the ground in response. Then we came to the *rufa* nest which I will label "H", some forty metres from the *polyctena*. Suddenly the Geiger needle jumped in an altogether abnormal way. We went some distance away and the counter returned to mere background noise. As soon as we came back to H it showed a considerable deviation again. No more doubt was possible. *Polyctena* had transferred isotopes to *rufa*. How? We do not know; we did

recall, however, that Gösswald of Würzburg observed something of the same sort among his experimental ant stocks. And,
what is more, it was between the same two species, *polyctena*
and *rufa* . . . However in that case the objection could be raised
that it was some deviation in behaviour caused by laboratory
conditions, but no such criticism could obtain in our case.
There must, then, be contacts between different species of ants,
other than those already known to us (namely war and slave-
making, of which more below).

THE TWIGS OF THE NEST

All my observations of the red ant have invariably and almost
forcibly led me back to its nest, a remembrance perhaps of the
many times I destroyed them with my foot when I was a child.
I do not plague the ants in that way any more: or rather the
torments I inflict on them nowadays are more scientific as well
as more subtle.

Let us take a look at this nest together: are you not struck
by a fact so obvious that we cannot see it? *It is clean*: by that I
mean that its surface carries only twiglets of almost equal size;
there are no oversize twigs, no dead leaves. It is, in fact, because
of this homogeneity of surface that we are able to recognize an
ant nest from afar. Yet dead leaves, for instance, fall all over
the wooded ground—look at the path we are walking on!
Therefore there must be some kind of cleaning system at work.
We shall see: for you must understand that a great many
experiments can be made in the depth of the forest without any
instrument other than branches, leaves, earth, stones and a
fair share of imagination.

For example, let us drop a dead or living leaf on top of the
ants' nest: considerable agitation ensues, the leaf is hauled by
its edge in a more or less unregulated way until it begins to
slide down the fairly steep slope of the mound and finally lands
at its base. How do the workers manage it? Is there
co-operation?

CO-OPERATION AMONG ANTS

I must now allow myself a digression which will, however,
quickly bring us back to the nest twigs. It happened several

years ago and the controversies that split the biologists of that time into two camps are no longer in fashion. Some were followers of Rabaud, a dogmatic and destructive character but at the same time a brilliant observer. Reacting against the excesses of an over-simplified finalism, he maintained, half-seriously, that there was no such thing as purpose in the living world: that the blind forces of evolution had haphazardly allotted to animals an incongruous set of senses and equipment with which they had to avoid death as best they might. For example, the social insects are by nature solitary; but a blind tropism, *inter-attraction*, brings them together without preventing each going about her work quite individually and having nothing to do with her companions. So much so that their mighty constructions are in reality nothing but illusions of the mind, however much we may long to perceive at least some co-ordination. The same goes for the transport of their prey: have you not watched, asks Rabaud, a number of ants carrying a dead insect into their nest? Is there anything more ridiculous? One pulls this way, the other that, each in a different direction from her neighbour. The mystery lies in the fact that their stores are nevertheless replenished, however senseless the means.

These corrosive arguments held their own during the thirties; they made us think more deeply. But the good Rabaud irritated me enormously all the same, and I had the feeling that he was altogether wrong: co-ordination and adaptation, might not be found where we had rather naïvely supposed without studying the phenomena with sufficient care, but did that have to mean that they did not exist, as Rabaud maintained after regarding them a little more closely? Should we not rather *look for them elsewhere*, as I hoped to show by examining the facts from a new angle? A pastime very much to the taste of a young research-worker—to prove that one of the Mighty Ones was talking rubbish. Here were motives enough to set me to work.

All the more since, in the fury of their argument, both sides had forgotten one very important point: to *define co-operation* and to measure it. The definition can only be empirical.

Supposing, for example, that two ants carrying a heavyish load progress more quickly than one at the same task: then we must conclude co-operation. *Not that it is the result of conscious planning as with man*: it may simply be a matter of one individual being stimulated by the presence of another, a phenomenon often observed among insects; and for that it is not even essential for the more or less benevolent assistance to make much difference or to be effectively co-ordinated as to direction. It is enough for it to be there.

There was an old pathway in my childhood garden which had always been much frequented by ants, and there I swept clear a length of about fifty centimetres and put down a marked ruler. Wasps were plentiful on the flowers that year and were just the prey I was looking for: of standard size and attractive meat for red ants. It is easy to kill them by pinching the thorax. When I offered the dead wasp sometimes to a single ant, sometimes to two, it was easy enough to confirm that transport was twice as fast in the latter case. And that should silence Rabaud!

The phenomenon is actually more complex than that, and I have described it in detail elsewhere. Let us stop there for the moment: in certain circumstances red ants do indeed show some sort of co-operation, whatever its exact mechanism may be: that is to say that the same task is accomplished more quickly when a number of individuals work together.

RETURN TO BUILDING

It is highly probable (though it is not easy to analyse this particular case) that ants co-operate in the upkeep and cleaning of the dome of twigs. Do they, then, give it constant attention? Certainly they do, but not only in order to keep it clean. It is far more curious than that. You can easily make the following amusing experiment yourself: delicately lift with tweezers a few twiglets from the mound (trying not to arouse the wrath of the workers, who are sensitive to the smallest vibration) and form with them a few letters on the nest surface. Then wait: it will not take long. An hour later three-quarters of your letters will have been dismantled. True enough, you may say, but it is probably only the result of the disturbance caused by moving the twigs. But there you are mistaken: another experiment will prove it. Let us fix a magnifying glass over one particular area of the mound, noting carefully the position of the principal twigs. I was forgetting that we are in the middle of a wood where no spy-glasses are available. Never mind, a piece of birch bark rolled up and tied together with grass, then fixed firmly on to a branch stuck in the ground, will do as well. This time we have touched nothing; we will merely view the nest through the birch-bark cylinder.

Every twig is moved: slowly (it can take twenty-four hours) but surely they are all displaced. Kloft, who was sceptical of this, proved it by an interesting experiment. He sprayed the surface of the dome generously with paint. The ants, offended by such treatment, covered up all the painted twigs with new ones. Up to here the story is not very extraordinary, since, as we shall see later, this is the normal reaction of the worker-ants to any strange object. For two or three weeks no change could be seen. Then the painted twigs began to reappear only to vanish again a few weeks later. This constant rearrangement of the twigs, these cycles, may be responsible for a curious fact: the twigs in an ants' nest never go mouldy, although they are often very damp; yet as soon as the ants leave, the twigs begin to mildew. There may also be antibiotic and anti-fungal substances in the nest which inhibit it from turning into a decaying mass. This has been maintained, but I do not know whether the evidence is as convincing as in the case of the beehive. In any case, according to the works of Pavan, the ants do possess at least one powerful antibiotic.

However that may be, their cleaning activities can be tested in a number of ways. What they particularly dislike is matches. I know that it is considered dangerous to handle these articles in woods, but those issued by the French Government are so difficult to light that the risk is greatly reduced. We can, for example, arrange the matches round the dome in concentric circles, which will enable us to measure the degree of the ants' activity while noting which matches are removed first and which last. The ants pick them up without difficulty and sometimes carry them to a fair distance; but they begin, it appears, at the top. This must be the most sensitive area, comparable, in the bee-hive, to the point of the comb. The experiment is even more striking if a handful of confetti is placed right on the summit. The pieces are all carried down, but always along a number of preferred paths which show that this dome, so uniform to our eyes, does not seem so to the ants.

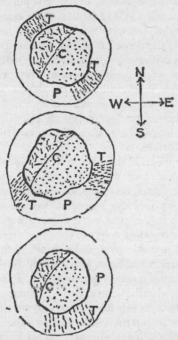

Fig. 17. Reconstruction of an ants' nest: 3 different nests showing the central stump laid bare. C, crescent of twigs; P, circular path swept and covered with paper; T, transport of the twigs to special sections of the track. (After Chauvin)

The matter has been fairly simple up till now, but let us advance further. Ants can easily pick up matches, but what would they do in the face of enormous poles, the size of a pencil for instance, firmly planted on top of the mound? What

I witnessed then I still cannot understand. Into the dome I stuck some little branches in the form of a crown surrounding the summit, and I waited; nothing very marked happened at first. It was obvious that the ants were very annoyed, for they scurried about and burrowed at the base of the branches, but these were too deeply planted and did not budge. Yet by the following week there were only one or two left in position and the others were lying at the foot of the mound. What had happened? Was it the result of chance and the constant rearrangement of the dome? If so, one could only imagine that the little stick would eventually lean to one side and that the ants would then undermine the dome at the foot of the branch until, in the end, it slid all the way down. This can only be supposition, since I did not witness it; and I dare not put forward a more complex psychic process to explain the resolution of this problem—which, after all, was one the worker-ants had never met before.

But what happens if the object is definitely too big and irremovable, like a chestnut leaf for example, unkindly nailed to the dome, by a stake through the middle? Now we discover something that is a veritable mania in ants and which we shall later see develop in the most extraordinary ways: *anything they cannot move they cover up with twigs*; so the leaf is promptly enclosed in the nest. Moreover, any hollow object seems to infuriate them particularly, for a large jam tin will be filled up with twigs in a few days. I amused myself by putting on the ant nest several large concentric cardboard cylinders about fifteen centimetres high. The ants filled them one after the other with twigs right up to the top till it looked like a Babylonian ziggurat in miniature. Who will explain such strange behaviour?

In any case it will help us to understand the technique employed to build the dome. It may seem exaggerated to use the word "technique" in reference to this vague heap of twigs, because, in spite of oneself, one imagines it being constructed by human methods. A man would empty a basketful of twigs on to the ground, and these would quite naturally fall into the shape of a regular cone. But this is not the way the ants do it. They carry each twig separately and place, not throw, it beside

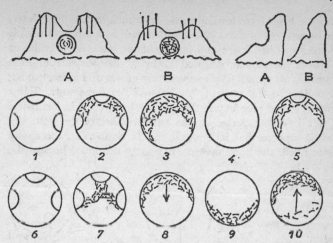

Fig. 18. Figures A and B show the filling up of a cavity, seen from the front and the flattening of two humps, seen in profile. Three little sticks completely imbedded in each hump were half-way uncovered (*right*). Figures AB (*right*) represent in outline how a hole is filled up from the bottom. 1, 2, 3, three phases in the levelling up with twigs of three peripheral humps; 4, 5, the same phenomenon with a single hump. 6, 7: ant nest exceptionally covered the three humps centrewise; 8 a crescent of twigs is transferred in the direction of the arrow, giving fig. 9; in 10 we see the reconstruction of the original crescent in its first position and the destruction of the crescent from which the material was drawn. (After Chauvin)

the others: in these conditions *there is no reason why the nest should assume the form of a regular cone* rather than some other more or less irregular shape. The only way to understand the process is to commit an even worse atrocity than the preceding ones; completely sweep away all the twigs and reveal the decaying old tree stump around which red ants always build their nests. When the first panic is over the brave workers begin once more to fight against a hostile universe. Once more they pile up twigs on the old stump, generally starting on the north-west side; not that they know their cardinal points but, I suppose, because in our part of the world that is the dampest side and therefore the side on which tree trunks decay soonest. The heap of twigs once formed seems to stimulate them (see later, in regard to termites, Grassé's theories on stigmergy) and it increases in size faster and faster. At the same time the horns of the crescent spread out and finally more or less surround the stump. Now the edges of the nest have grown up higher, leaving a hollow in the centre: "A hollow?" cry the ants. "How ghastly!" and give way to their mania by filling it in forthwith.

We have not yet explained the structure of the nest. We might suppose that its summit should be flat for, the cavity once filled, the ants have no reason to continue. But on the

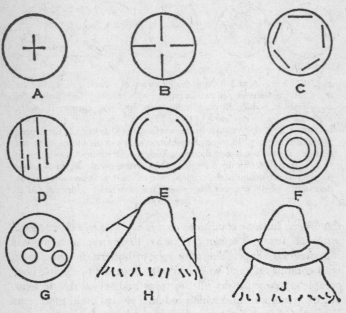

Fig. 19. Different positions of partitions on a *Formica rufa* nest: A, small cross-piece of slender branches; B, cross-piece leaving the centre free; C, partitions arranged as incomplete tangents; D, parallel partitions; E, circular partitions with a "gateway"; F, several concentric compartments; G, several hollow cylinders placed side by side on the dome; H, partitions forming an angle with the surface; J, partition in a discoidal ring form, encircling the dome halfway up. (After Chauvin)

contrary, the dome can still be quite pointed, although regular. This regularity can be explained by the ants' horror of anything concave. If, for instance, one seizes a handful of twigs from the side of the nest, leaving a fist-sized cavity, tremendous agitation ensues. The ants throw themselves into the hollow, which is rapidly filled in. This technique may appear somewhat violent, for the nest has a structure not always appreciated by the superficial observer. On the outside we find the finer, more closely laid material which doubtless gives more effective protection against rain; towards the centre the twigs are bigger with more space between, which allows the ants an easier passage. Therefore, if I seize a handful of twigs from the side of the nest I am removing the surface layer and laying bare the lower level which should not be exposed to the open air—a bad technique. So it is better to press the fist gently into the dome, which will yield without much difficulty, and thus avoid upsetting the organization too badly. Now we have a barely visible hollow. We must register its depth with precision; a forked branch will serve. Fix one end near the nest and arrange the other "prong" of the fork so that it just touches the bottom of the cavity. In a few hours we shall see that the dome has resumed its original shape and the end of the "prong" is no longer visible.

This brings us to another problem. An ant's sensitivity must be delicate indeed to register such slight irregularities. What kind of sensitivity can it be? I simply do not know, but I would be inclined to guess that unevenness affects the sense of balance rather than the eye. It should be added that a projection added to the mound takes much longer to level out than an indentation.

There still remains the question of the dome itself. Why a dome instead of a more or less level plateau? In an attempt to find out I stationed myself by an ant nest for several hours and watched the progress of 155 individual ants as they arrived at the nest carrying twigs. Where were they making for? Their routes on the mound were so capricious, it was hard to tell. Some dropped their loads almost immediately, while others wandered about for twenty minutes before deciding. However,

I would estimate that 50 per cent placed their twigs right on or near the top, and 50 per cent left theirs on the circumference. Since the summit is of course smaller than the periphery, it becomes clear that the relatively greater number of twigs

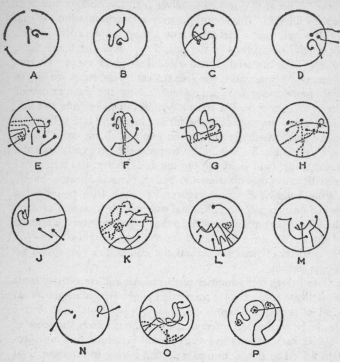

Fig. 20. Individual tracks of ants climbing the dome with twigs. The dotted lines correspond to the transport of particularly heavy loads. (After Chauvin)

unloaded on the highest part of the centre give the dome its conical shape.

We're nearly there, after arduous reasoning and very simple experiments, which, however, gave us a great deal to think about. There are still one or two points that we do not really understand. For instance, if I put on the summit a cruciform partition made of four small boards, the heap of twigs which will pile up round it will always be larger in one of the quarters. If I take the partition away to replace it a little later, the preferred quarter may turn out to be a different one this time. Why on earth do the ants not bring equal contributions to all the quarters? On the other hand, if the partitions are imbedded

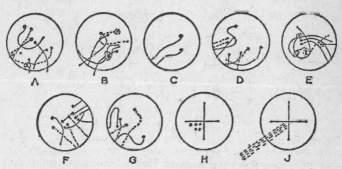

Fig. 21. A continuation of Fig. 20, except for H and J. These two last figures symbolize my hypotheses on nest construction. I first supposed that when a cross-piece made of thick branches was placed on the mound the more active ants found themselves to some extent imprisoned in one or other of the quarters and hardly left it; this is represented in fig. H and would account for the mass of twigs being greater in certain sections. In J the other, more probable, hypothesis is shown: the ants would not be forcibly confined by the partitions, but the contributions of building material would not be equally plentiful from all quarters. The dotted lines indicate the direction from which the more abundant contributions are coming, and the partitions prevent its equal distribution over the whole surface of the summit. (After Chauvin)

equidistantly on the sloping sides of the dome, converging towards the summit, but leaving it free, the mound keeps its symmetry. . . . A nest of wood ants, as we see, conceals more than one scientific problem.

We must not leave our builders without emphasizing their ogre-like appetites. The Germans, patient observers par excellence, have counted hour by hour all the insects that the foragers drag into the depths of the nest to be devoured: they add up to nearly *one kilo a day*. However, the principal occupa-

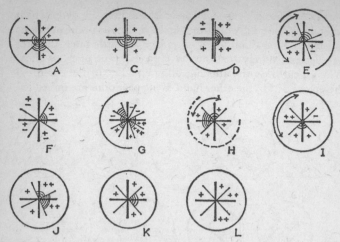

Fig. 22. Summary of experiments with partitions on *Formica rufa* nests. The thick cross represents the four cardinal points, the thin cross the direction of the partitions placed on top of the dome. The small crosses seen between the arms of the big ones represent the varying degrees of activity displayed by the ants in different sections. The parallel circular sectors which join the arms of the cross in some of the diagrams correspond to the building activity: there are more or less of them in any particular quarter according to the amount of this activity. The circles or semicircles round the diagrams show the degree of shade: some ant nests were deep in the forest where no daylight showed between the tree-trunks. Nests E and I include an arc ending with arrows: these represent the "cyclosis", the rapid running about round the dome that the ants sometimes excute after the partitions are placed there. It is clear that in this experiment the building activity is dependent neither on shade, nor on orientation, nor on the running-about of ants on the surface of the nest.

tion of the red ants is climbing the fir trees in order to suck honey-dew, in other words the excrement of the aphids; the ant colony consumes nearly a hundred kilos of this honeydew every year. It must be remembered that the population of the biggest nests can be over three or four million.

Recently the ants' predacious instincts have begun to be exploited commercially. Gösswald of Würzburg and Pavan in Italy noticed that as long as woods were well enough stocked with ants' nests, harmful forest pests were held in check! Soon forests that were short of ants were being supplied with them —to the extent that there is now a flourishing trade in ants between Austria and Italy; the former sends the latter ant nests by the lorry-load. . . .

Fig. 23. Two worker ants regurgitating and exchanging food; the soliciting ant is on the right. (After Berlese, in Grassé)

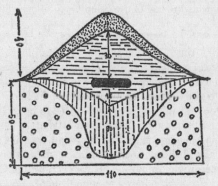

Fig. 24. A red ants' nest after Raignier. The scale is in centimetres. The surface layer made of finer material is indicated by dots; the coarser central material by horizontal dashes; the rotten central stump in black; the vertical dashes indicate the decayed material.

ANTS IN WINTER

But what happens to the ants during the cold season when the dome is completely deserted? It has not long been known: in fact they dig a hibernation chamber more than a metre below ground level where the temperature is constant at about 10°C, and here they mass themselves into a great ball until the weather grows more clement. But how can they tell when it is time to resume their activities? The Germans found the answer quite recently: it was a brilliant piece of work, the discovery of the "thermic messengers". Certain of the workers are less sensitive to cold than others and wander almost ceaselessly between the hibernation chamber and the surface of the nest. When the temperature rises and a ray of sun allows it they come up to get warm, then go down again bringing with them a few extra calories to warm the chamber. Gradually, as the sun grows hotter, more and more ants reach the level of activity, roused by the ever more excited messengers, until at last the temperature of the hibernation chamber reaches a point high enough for its whole population to go above ground and take up once again the everlasting toil of the colony.

THE HUNTRESSES

It was the first time I had ever set foot in tropical Africa, near Abidjan. I still remember my astonishment when I left the air-conditioned atmosphere of the ship and was suddenly plunged into the furnace of Treichville and Adiopodumé, via Abidjan. One tells oneself it is not possible to be so hot, it must stop ... then gradually one becomes a little more used to it. ... Sitting in the passenger's seat of the jeep I was not so stunned by the heat that I didn't notice the black snakes that lay across the road, seemingly immobile. How was it that Gaston, the black chauffeur, did not try to avoid them? For I had been warned: before becoming a Christian Gaston had been (or still was) a witch doctor of the Ebrié tribe and his soul was in the snakes. If he does not hurt them, they will not do him any harm; consequently he can go hunting alone in the forest at night and return intact—at least he has up to now. Incredible!

It is also true that in his role of sorcerer he must have killed and eaten at least five people, since that is part of the initiation ceremony. He is not very talkative on this point, but he is always careful to stress that he has, at any rate, never eaten a white man, "because they are counted"! Though the best of fellows, he does have the foible of taking corners on the hub-caps. Then why the devil does he run over his totem without giving a rap? At that moment he must have intercepted my glance from himself to the road. "It's only ants, boss," he explained.

Ants! These can only be the famous army ants of which I have so often heard, the huntresses before whom all the beasts of the forest draw back in fear, even the largest. I knew a little about them, without ever having seen any. They are very common in tropical Africa; terrible stories are told of them, particularly of the method used by Negro kings to dispose of courtesans who had ceased to please them. They tied them, still alive, across the path of the ants. A few days later only the well-scoured skeleton remained. One of my friends, the director of the Zurich zoo, also told me of a personal experience that left him sorely discomfited. He is interested only in large animals, over 500 kilos, for his zoo. Thus, when he made his first African expedition and heard about the army ants, he was rather sceptical and chalked it up as one of those blood-curdling tales kept for greenhorns setting foot in virgin jungle for the first time. Thereupon his expedition set out and he stopped in the bush to spend his first night under canvas. The heat, as usual, was stifling, so he went to sleep naked under his mosquito net. In the middle of the night a strange sensation woke him up: by the light of his oil lamp, which he had forgotten to put out, he saw a nightmare vision: his whole body, his bed, the net (which must have had a hole in it), all were black with crawling ants which, as soon as he moved, began to bite him. Agony! My friend leapt out of bed and pulled on his canvas boots: they were full to the top with ants which immediately covered his feet with blood. "I sprang out of my boots," he continued, "driven wild by the ants. I jumped on to a paraffin tin which happened to be there. The noise woke up the blacks, blast them, and they simply rocked with laughter to see their master yelling, naked, on his pedestal. Then they did the only thing they could: soaked me with paraffin, which the ants hate. I was never so scared in my life!"

Here I am beside a marching column, keeping beyond the critical distance. There is nothing to fear as long as one leaves 20 to 30 centimetres between one's feet and the edge of the procession. If one neglects this precaution, the workers often swing round and attack one in the rear. It is three o'clock in the afternoon and according to what I am told by Birhama, my African assistant, they have been passing by ever since eight o'clock this morning. The workers are hardly bigger than our red wood ants; they advance at about the speed of a man at the double, forming a column about an inch wide. The sand, worn down by millions of little feet since morning, has now formed a fairly deep gully along which they march. At the sides of the gully are scattered the soldiers, with their formidable mandibles, shielding their people below. One party of workers carries the young larvae and all stages of the brood. Another party goes hunting: they climb a tree and occupy it completely, down to the smallest twig. All the inhabitants of the leaves and branches retreat before these inexorable aggressors and finally drop to the ground, where more ants are waiting to tear them to pieces. Observing more closely, it appears that not all the forest's inhabitants are afraid of ants after all, for I can distinguish, flying at a fixed distance above the procession, a cloud of minuscule flies which look to me like members of the Syrphid family. Every now and again one of them darts down like lightning on to a worker and immediately afterwards flies up again. It is not easy to see, but it seems to me that it had glued an egg on to the ant's back. Thus it would appear that even the most fearsome ants have parasites and predators, like every other creature alive.

These ants have no fixed nest; they make only temporary "camps", great globes of massed workers which, however, allow other ants to pass through to the centre of the mass, where they move about. Moreover, Schneirla of New York has recently shown that the raids are stimulated by the queen's oestrous cycle: in other words, this cycle includes successive bursts of egg-laying, spaced out in time, and it is on these occasions that the ants suddenly increase their activity and set out on their devastating raids. Sinister denizens of the forest!

THE WEAVER ANTS

It has been known for the Negroes to refuse to pick coffee berries in a plantation because of another species of ant: the

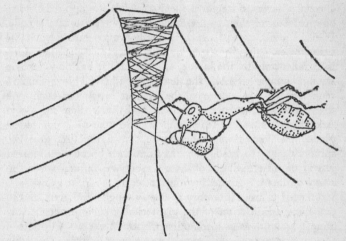

Fig. 25. A weaver ant (*Oecophylla*) holding in her mandibles a weaving larva, which she uses as a shuttle. (After Bugnion)

Oecophylla, or weaver ant, which haunts the branches and drops on them from above, biting them fiercely. These weaver ants furnish one of the most curious sights among the ant tribes on the Ivory Coast. In the coffee plantations one sees bunches of three or four leaves joined together by silken strands. Never thinking of *Oecophylla*, I supposed at first that they were spiders' nests; but when I opened them I was delighted, in spite of the many ant bites on my hands, to discover weaver ants inside.

I spent a whole Sunday sitting by a coffee-tree, positively

hypnotized. With finger and thumb I prised two of the nest leaves wide apart; some guards emerged in fury, but by this time I had learnt to keep my fingers out of their way. Very soon they yielded place to chains of ants who, by clinging on to each other, succeeded in joining the two leaf-edges together again. One, two, three chains took up their positions, the workers flexed their muscles, and lo and behold! the leaves were drawn together. From between them crept an ant whose intentions were obviously quite different. In her jaws she clasped a larva of her own species, which she applied first to one leaf border then to the other, using the grub as a shuttle. It emits a silk thread which is finally woven into a veritable tissue, after which the "traction-ants" leave, for the silk (which is secreted only by the larvae, not the adults) has now become strong enough to hold the leaves together. Thirty times I reopened the gap, destroyed the weavers' work and even took the "traction ants" out of the nest and threw them far away. And thirty times new teams appeared to take their comrades' places, the leaves were once more drawn together and the spinners came to bind them. My patience became exhausted just as the brief twilight of those regions was falling, when the noises of the bush suddenly grow louder as the light grows less. I returned to my laboratory in deep thought, followed by the deafening stridulation of the giant cicadas whose chorus was almost as loud as a locomotive whistle. Insensitive robots—that's what ants were; I must have almost emptied the nests by carrying off so many workers. Yet always one more team arrived imperturbably to repair the breach.

I gave myself up to numerous experiments until my hands were so badly bitten that the pain forced me to stop. It must be admitted that my methods had been somewhat brutal and that the bites were not entirely undeserved. My procedure was to transport a complete nest of *Oecophylla* to my laboratory in a glass case. I opened it quickly and, with the back of my hand, swept all the ants out of it. Then I gave them one or several coffee leaves in order to study their method of repairing the damage. After a short interval of utter panic groups began to form near the scattered brood; gradually the smaller groups merged into the larger, as is the way of social insects in general (see above, Lecomte on the Bee). Everyone climbed on to the leaves, carrying the brood, and almost immediately the traction ants were at work.

"Nothing simpler," the ingenious observer may object, "the *Oecophylla* will tug at anything you offer them, or that is within their reach, so naturally they tug at the leaves." It is true that if you present these workers with a twig, they will bite and pull it a bit; but if it is a leaf, or leaf-sized object, their behaviour is quite different: they exert themselves desperately, try to hook on to it with their mandibles and climb on to it if possible. To begin with, then, ants do not pull in the same way at anything, but are especially interested in leaves. Do they pull at random, as was suggested? At first sight one might think so, for one sees them seize the leaf-edge from all directions, parallel with it or at right angles and so on. Despite this, in very few minutes the leaf is folded and fastened in such a way that the brood and the serried ranks of nurses surrounding it are all enclosed; never have I seen a leaf folded inside out. What complicates our interpretation is that ants do not set about it as humans would: it seems that when a nest must be rebuilt a general urge to pull develops, and some of the workers react to it by tugging the nearest object in any direction. Nevertheless, it is obvious that the efforts *of the majority* are correctly orientated, for the folding of the leaf is accomplished in a few minutes. Moreover the leaf is drawn only by its edge, which appears to be a very sensitive zone. The workers could as easily get hold of a thread attached to it or of the body of an ant already pulling. In short, the traction is governed by a whole set of social regulations of a superior order which adapt it to one purpose, the protection of the brood. We shall be wasting our time, however, if we try to account for everything by a system of simple reflexes: the body social knows how to adapt to a situation (one-leaf or two-leaf nests, for example) and organize the workers in the most suitable way. We see the same phenomenon in the transport of prey where, *though all appears to be in disorder, the dead prey is conveyed rapidly to the nest*: only preconceived ideas could cause such an important fact to be overlooked!

HARVESTING, STOCK-RAISING AND SLAVE-MAKING ANTS

Ants do not confine themselves passively to sucking honey-dew from the aphids: they actively protect them from their enemies, such as ladybird larvae. This had been suspected and recent English works now seem to have solidly established it. In addition, when the cold season approaches, the ants collect the eggs, the form in which aphids overwinter, and bring them into the nest, where they will pass the cold months safe from any possible enemies. When spring comes and the winter eggs hatch, the ants carry the young aphids on to their pasture plants. They take great care, however, to bring them in again every evening as long as the nights are liable to turn cold. Only gradually do the ants allow their cattle to graze permanently on the plants and even then they plant watchful sentinels to guard over them. Some species cultivate root aphids and surround them with veritable miniature stables, carefully fashioned in earth, that prevent any escape.

Other ant species are interested in grain: these are the famous *Messor* spp., the ants mentioned in the Bible, in the Book of Proverbs: "Go to the ant, thou sluggard; consider her ways and be wise; which . . . gathereth her food in the harvest." This harvest is none other than the "ant corn", those litres of grain harvested by the workers to feed their larvae. There are many legends about them and unfortunately it is a long time since a really experienced myrmecologist has observed *Messor* from close to: it has been said, for instance, that the large soldiers of this species grind the grains to a paste with their great mandibles and put it out to dry—like leather!—in the sun before giving it to the larvae. More probably it is simply stores deteriorated by damp that the soldiers put out in the sun.

Then again, cereals of the same kind are often seen to be growing in a ring round the *Messor* nests, so of course it is said that these ants are cultivating the grain, whereas it is far more likely that some of the less careful workers have dropped a few grains which have subsequently germinated. It must be added that when the garnering fever possesses the harvesters they will even gather up morsels of coal or little bits of glass, though afterwards there seems to be some sorting out of anything inedible, together with grains rejected for some unknown reason.

If I do not appear more excited by the hypothesis of cereal-growing ants, it is because cultivation is indeed practised by certain other species, using methods far more complicated than those required for cereals: I speak of the fungus-growing *Atta*.

Fig. 26. The honey ant, *Myrmecocystus. Left*, a reservoir ant; *right*, a normal worker. (After Picard)

This large, South American species, called parasol ants by the natives because of their habit of walking with a piece of leaf held over their heads, invade orchards, which they devastate by cutting the leaves of the trees. The leaves are carried back to the nest, hacked into small pieces and sown with a single species of fungus, the mycelium of which serves to feed the larvae. The enormous mushroom caves of the *Atta* stretch for many metres underneath the earth and paths: in short, it is a real scourge. Each species has its own kind of fungus. When the young queen is ready to be mated she does not forget to take with her some fungus mycelium in a special pocket of her mouth cavity. After her fertilization she digs a little chamber, where her first care will be to reconstitute the fungus beds. To this end she crushes the first eggs she lays and regurgitates the

Fig. 27. The fungus-growing ant carrying leaves. (After Goetsch)

mycelium on to them. Only when this sprouts does she allow her young to hatch out, and their principal work will consist in supplying the bed with freshly-cut leaves.

THE EXPLOITATION OF OTHERS

Though ants may follow the law of relentless and productive work, they can also indulge in other activities which, by human moral standards, appear less respectable. Let us begin by looking, for instance, at the founding of a colony. The simplest cases are similar to that of the *Atta*, where the fertilized queen manages all by herself. With some other species, however, she is incapable of such independence and must have help. Very well, let her fetch some of her own kind, you may say. This is what often happens, but some queens are less particular and do not object to associating with workers of another species. Obviously the workers will die after a certain time, but they will have had time to bring up the queen's first brood. A black wood ant with an ether-like smell, the *Dendrolasius*, cannot even begin to found her colony without the help of another species. Thus far we have considered only cases where there is mutual consent, if I may so describe it, or at any rate no violence. But the queen of a rather rare species, *Anergates*, simply penetrates into the nest of another race, the *Tetramorium*, and, after a time, some strange perversion of instinct makes the workers murder their own queen to make room for the interloper. In doing so they condemn their mother colony to death, but the *Anergates* queen will have had time to procreate a new generation of her own species.

Ants have also invented *slavery*. The *Polyergus*, for example, are practically incapable of any activity except waging war. Their unfortunate disposition has been attributed to the shape of their long, sharp mandibles, typical weapons of war, it is said, too fine to be suited to the domestic tasks for which the short, stout jaws of the *Formica fusca* are so well adapted. Theories of this kind, based on morphology, are somewhat naïve and for my part I do not believe them. The world is full of species carrying out similar tasks with different tools, or the same task with very different equipment. But it is some curious and as yet unplumbed quality of their psychology that deter-

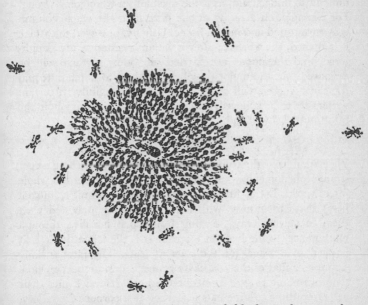

Fig. 28. The female of *Myrmica rubra* surrounded by her workers; note the "sentinels" round the outside. (After Goetsch)

mines the *Polyergus'* behaviour, and that is that. Their expeditions always take place during the hot hours of the afternoon, and I will borrow the description of the Belgian Jesuit, Father Raignier, the great ant specialist: "About three o'clock in the afternoon, a few amazons run excitedly to the surface of the nest, crowd together and mutually beat each other's head and thorax with sharp strokes of the antennae until enough soldiers have poured out of the nest. All of a sudden this seething mass forms itself into a column twenty centimetres wide and advances resolutely, with no previous consultation, yet no hesitation as to direction or route, towards the colony to be pillaged. The vanguard hastens onwards while fresh waves of soldiers are still emerging without pause from the nest opening; no auxiliary worker is among the troops." This gives a strong impression of co-ordination. But sometimes manœuvres follow that are difficult to understand, as if the mechanism were going wrong. For example, on arriving at the *fusca* nest, the whole column may turn round and march home. Usually, however, the attack is launched, the workers, after a valiant resistance, are slaughtered, and the pupae are carried off. Some of these will be devoured, but the majority will be allowed to hatch out; and the slave-workers will serve their owners faithfully. If certain writers are to be believed, some species of slaves even join in the raids on their own kind.

A SOCIAL PERVERSION: "LOMECHUSAMANIA"

We cannot leave the world of the ants without describing a very strange phenomenon, without parallel, perhaps, in the whole insect world. Let us first note that ants (and termites, but not bees) share their nests with a host of commensals and treat them in various ways ranging from open hostility, through indifference, to the liveliest interest. It is the last attitude that has led to disaster, for the ants at least. The most striking example is that of the *Lomechusa* genus, a minute beetle whose sonorous name rings a knell for the ants. It crawls into their nest, from time to time eats a worker at a corner of a gallery, lays its eggs amidst the brood and lets them be fed by the worker ants, who apparently see nothing amiss, although the

young *Lomechusa* grub has an enormous appetite and is not
above devouring one or two of the young ant larvae around it
in addition to the food it is given. If by any chance an ant does
try to chase or attack the beetle, the latter raises its hindquarters
and displays the *trichomes*, humid hairs that the ant licks with
relish, and which secrete a deadly poison. The sturdy workers
are lost and their nest with them. They forget the superb
machine of which they were the driving force, their tiny,
terrible world, the thousand tasks about which they should be
busy; from now on, nothing exists for them but the cursed
trichomes that distil oblivion of their duty, and death. Before
long they lose their equilibrium and can no longer move along
the subterranean passages; their undernourished larvae develop
into deformed adults. Soon the nest will dwindle and disappear.
And the *Lomechusa* will go in search of fresh victims in a neighbouring colony.

It is on this grim note that we will leave the world of the ant:
a splendid machine as strange to us as if it had fallen from the
planet Mars. And now let us visit the domain of the wasps and
the subterranean country of the termites.

WASPS

If we happened to have some of the magnificent *Polybia* of the
Amazon in the laboratory, our researches would be much
easier! The enormous nests of these wasps are as high as a man,
and are fixed to a branch by a thick loop. *Polybia* gather honey
but it seems that at the same time they collect prey, which they
bring into the nest; perhaps the captives are not dead but
paralysed so that they should not decay—the larvae of many
hymenoptera feed on the bodies of live prey which their
mothers have paralysed by means of judiciously placed stings.
We cannot be sure, for the *Polybia* are to be found only in somewhat inhospitable regions, and what we know of them would
not fill more than half a page.

Wasps can hardly be called rare in our part of the world,
but their behaviour seems to be much simpler. Nevertheless,
there are interesting data to be collected. Let us begin with the
development of the colony; Deleurance studied this with great

care in the case of a primitive species, the *Polistes*, whose combs are open to the sky, unprotected by any cover. This wasp tears off wood from here and there, chews it and turns it into a sort of paper; from this the cells are formed. Most wasps do this and it was Réaumur, at a time when paper used to be made from old rags, who remarked very appositely that we should imitate the wasps and use wood pulp.

With the *Polistes*, building activity is periodic and is repeated several times a day; in fact the nest never ceases developing and can never be considered definitely complete. The immediate cause of the activity is the presence of eggs in the ovary; but everything depends on a discordance between the number of eggs laid and the number of cells in the nest. When the brood hatches the wasps feed the very young larvae on scrambled eggs taken from other cells. This goes on until there is a certain number of empty cells, and then building activity is inhibited. As the larva grows older its diet changes: the nurses feed it with animal and vegetable juices and no longer with the crushed eggs. Since the queen continues laying there comes a moment when there are very few or no empty cells left, and then the building begins again. Note that these nests, unlike the bee-hives, last only one season; only the queens overwinter, alone and hidden under tree bark or in nooks and crannies in the trunk. At the end of the season a special comb, the abortive brood comb, appears, which the wasps will sacrifice before it is ready. Far from stimulating building, it seems rather to impel them to destroy and desert the nest; furthermore, if the half-built nest is brushed with extract of abortive comb, it will be in danger of destruction, even by the queen herself. *Polistes* wasps, unlike bees, though they will often repair the edges of broken cells, will never, or only very imperfectly, stop up a hole in the cell wall. According to Deleurance there is also no division of labour such as bees practise. Pardi observed a *dominance*: that is, certain individuals pushed the others about and specialized in egg-laying, while the other females confined themselves to collecting food and building materials but did not lay.

When studying the construction of insect nests (and perhaps even when studying insect behaviour in general) it is inadvisable to consider only the simplest examples, such as the really rudimentary nest of the *Polistes*. Certain complex acts can be perceived only dimly with a very primitive species, and it is difficult to pinpoint them, whereas with more highly evolved wasps these activities are fully developed and easily recognizable. This was Vuillaume's idea when he began to study in my laboratory the wasps of our own countryside, whose nest, protected by a wood-pulp paper covering, hangs from a branch or is buried under the earth. His scheme for obtaining the wasps was ingenious: a small advertisement in the local press announced to the inhabitants of the Chevreuse valley that a simple telephone call would bring distinguished specialists to relieve them of their wasp nests. The response was amazing: we received three or four calls a day for several months. In the end we hardly knew where to put all the wasps, especially as they were beginning to attack the laboratory hives in order to steal the honey. However, apart from their aggressiveness, which is considerable, particularly in the ones with subterranean nests, they are creatures of absorbing interest and, in quite unforseen circumstances, gave us the surprise of our lives. Judge for yourselves.

The first problem in regard to the subterranean wasps was to get them to live outside their cavern, where observation was so difficult. At first Vuillaume and his pupils had no idea how to set about it, though one of the nests hung on a branch in the open air continued to develop fairly normally. By some chance another nest was deposited on the ground and forgotten for several days. At the end of that time we noticed it and stopped in our tracks with astonishment. We remembered perfectly well that it had been twice as big as a man's head and that we had left it on fairly hard ground. How then did it come to be half-buried, with a cap of earth on the top? At first we thought of some small boy's prank, but a nearer look showed us that the true explanation was stranger. The wasps were responsible for this amazing feat, for we soon saw them at it.

On reflection, there is nothing so extraordinary in the nest being buried. For this is how the matter proceeds underground: the queen looks for a mole's or field mouse's burrow and then, in the most spacious part, she attaches to the ceiling the first cells in which the eggs will be laid. Soon the nest has grown so much that it reaches the floor of the burrow. Then the workers take some earth away so that it has room to extend still further,

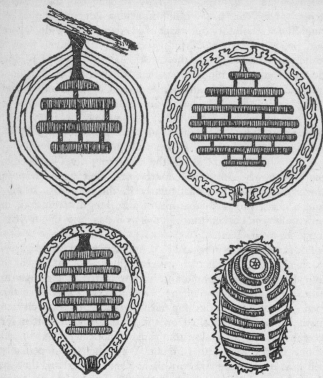

Fig. 29. Different types of wasp nests.

and they carry the pellets of excavated material outside, *but never, absolutely never,* do they deposit it on the upper part of the nest, which always remains free and clean inside it's cave.

What does the curious manœuvre of covering the nest with a cap of earth mean? Can it be intended as a protection against the rays of the sun? But the nest has never before been exposed to the sun: how can we believe that these insects could invent—and invent so quickly—a means of protection? Be that as it may, Naulleau tells us something more: if the nest, still on the ground is covered with an opaque box, the workers stop putting earth on the top, though they continue to bury the nest. If, on the other hand, the box has one glass side, the earth-cap will begin to form again. Thus heat and light—more probably light—are what govern the wasps' strange behaviour. But has this behaviour, in fact, a protective function? And is light harmful to the development of the larvae or any other element in wasp life? We simply do not know.

However, as we have seen, the workers always bury the nest, whether it is in the light or not. Here again the wasps had a surprise in store for us. When Vuillaume realized what they were trying to do he was dumbfounded, and hurried to pour out his story to me. "Since they are so keen on burying it," he had thought to himself that morning, "the experimenter (or rather tormentor) had better try to prevent them." Thereupon he slid a large sheet of glass beneath the nest. It had no noticeable effect during the first day, but after a few days the nest began, bit by bit, to change its shape; a little time later the nature of this deformation became clear, unbelievable though it may sound: *the nest was gradually shifting towards the edge of the glass*, rebuilt, millimetre by millimetre, from the inside. I realize that it is hard to credit this, and the reader may well accuse me of exaggeration. Nevertheless, at the end of the week it had reached the edge of the glass panel and had already thrown out an extension which was beginning to sink into the earth. It is all perfectly true, and we saw the same thing happen on several occasions afterwards. It can, of course, happen in nature that a large stone hinders the growth of the nest, which will then have to re-shape itself in order to continue growing. This may explain to some extent what happened on the glass panel, but we must remember that this was no normal growth of a nest, but rather a displacement by means of reconstruction; the nest stayed more or less the same size from the beginning to the end of the process. We can not say any more at present as we have not yet completed our investigation.

All this serves to point to a conclusion which has been slowly forced on me in the course of our researches on animal behaviour. The principal difficulty lies in putting your question in a language intelligible to the subject: thus, for a spider we must put it in terms of web; for an ant in terms of twigs; for a wasp in wood pulp; for a termite in grains of earth and for a bee in wax. It is a very simple truth, but one often neglected by investigators.

TERMITES

These insects constitute a challenge to an over-simplified view of evolution. The termites, morphologically archaic yet with complex instincts, are of very ancient stock. They first appeared in an indeterminate period; probably soon after the cockroaches, at least three hundred million years ago. Thus they came infinitely earlier than the bees and ants, *and yet their social machine is just as complex*. The complicated social patterns of the termite as we know it today must have evolved over a period stretching back into the incredibly remote past; but of that evolution no traces are to be found. At any rate, like the ant, it is known only in the social state; no one has heard of solitary termites. It must also be noted that its organism is rather simple and primitive, resembling that of the cockroaches, which is one of the most ancient insects. But, as with the other social insects, *as long as termites are interconnected*, their social instincts are in no way inferior to those of ants or bees.

It will perhaps be advisable to describe something of their habits; for though the public may know, or think they know, ants and bees which they see every day, they are necessarily far more ignorant in regard to termites. They are tiny white creatures (only the sexual forms are coloured) who, as a general rule, detest light. For this reason they construct nests out of earth, some of which are enormous; Grassé reports one in Africa *more than a hundred metres* in diameter, upon which the Negroes had built a village. To reach their food, which generally consists of dead wood, they build tunnels of earth. The wood is digested by a unique process. It must be noted that termites are no more capable of digesting wood and transforming it into food than the rest of us: but in their intestines they harbour an extraordinary fauna of infusoria who do it for them. After that the termite has only to use the products of these symbionts' digestion, or to consume the infusoria themselves! All wood-eating animals harbour such guests or they would be unable to live: this has been verified by research-workers who found a way of killing the intestinal fauna without harming the termite. The insect continued to eat wood, but died of starvation in a very short time.

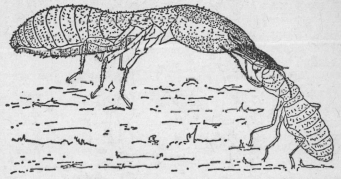

Fig. 30. A small *Bellicositermes natalensis* worker feeding a big soldier termite. (After Grassé)

The termites have yet another use for wood that is, I believe, unique in the insect world; they, like the ants of which I wrote above, cultivate fungi, but use them for a different purpose. A termite fungus cake resembles a small blackish sponge, very wet; they are numerous in the cavities of the nest. The fungus grows on this cake, which is made of finely minced wood. For a long time it was supposed that the purpose of the fungus was to predigest the wood by splitting up the cellulose and turning it into digestible sugar, a device often employed by other xylophagous insects. Grassé and Noirot, however, proved that this particular fungus has more unusual properties: it is not so much the cellulose but the lignin that it can split up into usable fragments. This is very surprising, for lignin is the hard, brown

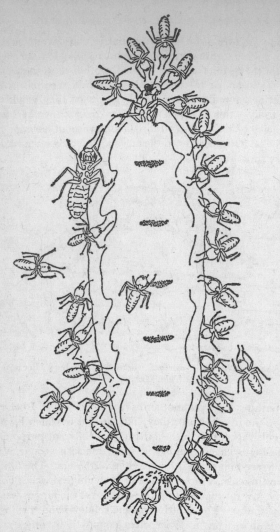

Fig. 31. The enormous termite queen surrounded by her workers (*Bellicositermes* sp.); *at the bottom* of the page workers are collecting the eggs; *at the top* other workers are feeding the queen; *top left* is the male. (A simplified sketch after Grassé. The scale is incorrect: in reality the workers are relatively even smaller.)

material that permeates the structure of wood cellulose and is far more stable and less liable to deteriorate than the latter. Up till now the most advanced studies of insect feeding have failed to show a consumption of lignin as opposed to cellulose and other constituents of wood. Only the termites' fungus can perform that feat. The insects gnaw the oldest parts of the fungus bed, where the lignin has disintegrated, then add new wood on top, which the fungus can attack at leisure. They do not, however, eat the fungus itself as do the ants.

In a favourable environment the termite population reaches an incredible figure. Grassé says that in tropical Africa it is almost impossible to break the ground with a pick anywhere without unearthing termites. Their unceasing movement of the soil and subsoil must undoubtedly have a great effect on humus formation. The almost universal presence of these insects can be explained by the immense fecundity of their queen. In some species the queen is the size of a sausage while the male and female workers (for Noirot has recently shown that the two sexes can be distinguished among this caste) are only as big as ants. She lays hundreds of eggs *a minute*, and in the vaulted royal chamber in the heart of the termitary a crowd of attendants wait on her, lick her, even bite her occasionally in order to drink avidly of her blood. Others progress round her, always in the same direction. It is a strange merry-go-round, where some give the queen food at one end, while others carry off the eggs, laid in machine-gun tempo, at the other extremity. During this time the male, smaller than the queen, but still huge compared to the workers, remains there almost unmoving. He does not die, as with other social insects, but survives in the royal chamber to re-fertilize the queen from time to time.

THE NEST AND THE THEORY OF STIGMERGY

The habits of termites are no less fascinating than those of ants, but it would be impossible to deal with them even summarily in a book of this size. Let us concentrate, therefore, on one of their most striking works, their nest. It is my opinion that even those of ants, wasps and bees cannot compare with it for perfection and architectural complexity. A Belgian scientist, Dr Desneux, has spent his life analysing the structure of African termitaries, and his illustrations of them leave one amazed. One would never think that they were not the work of man. They resemble banks or mounds of which the inner walls are in the form of spiralling colonnades with a complex system of passages running side by side, passing one below the other, intertwining, and all as regular as if made by machine. We still know nothing of the use of these arrangements, partly because the workers have seldom been caught in the act. Nor do we really know how they set about their task.

But the problem does not lie there; the eternal question still tantalizes us: how can these minute creatures build these St Peters, these pyramids, without a plan? Must we locate it in the brain, or invent one of those cosy substitutes such as the mind of the hive or of the termitary? To start with, it is very difficult to credit such a plan to any individual insect brain, for if there is one thing certain it is the stupidity of bees, ants and termites *taken individually*. Isolated, they are incapable of anything but dying in a very short time, for reasons that are little understood. Nor do matters greatly improve when they are together in small groups. We have seen in the case of bees how none of the functions of their social life can be carried out unless there is a certain minimum of participants. It was with this in mind that I postulated the theory of an interconnection of individual brains, based essentially on a comparison with calculating machines.

Grassé postulates another theory which unquestionably accounts for a certain number of the facts. I do not believe that it can explain them all. Firstly, three stages are to be distinguished in the behaviour of a handful of termites taken from their nest and put into a basin with some building materials. The first stage, called *unco-ordination*, is easily explained by the disturbance caused by these unkind manœuvres. They run here and there distractedly, and only after quite a long time do a few begin to work. Then comes the stage of *unco-ordinated work*, in which work grows more intense but is purely individual; here the termites conform to Rabaud's theory which, as we have seen, postulates quite definitely that social insects are held together only by blind instinct and that each busies itself with its own task regardless of its comrades. At this point the pellets of earth or wood pulp are carried and placed quite haphazardly; the beginnings of a passage may be roughed out, but the workers are quite indifferent to each other's contribution and often a pellet put in place by one termite is immediately removed by another. Because of this, progress at this stage is quite arbitrary. It is the same with bees newly introduced into an empty box: they lose no time in sticking specks of wax on the ceiling, but quite at random.

After this comes the stage of *co-ordinated work*. At a certain point it can happen that two or three pellets are stuck one on top of the other purely by chance. This proves to be a great incitement to the other termites, stimulating them far more than a single pellet. They bring new material and build it up into a column. When this has attained a certain height they stop putting their grains of earth quite on top of the pillar and build out at a slight angle to form the spring of an arch. Then, however, work stops, at least temporarily, unless the builders find another arch, or part of one, near enough to link up with. Grassé has also pointed out that the blind *Bellicositermes* workers (who in any case operate in total darkness) can join up an arch quite accurately, without having seen or touched the opposite curve. It is hard to tell how they can sense the proximity of their fellow-builders. Grassé was inclined to adopt Forel's "topo-chemical sense of smell" hypothesis. That writer thought that ants could distinguish the "long-shaped smell' of a twig, the "rounded smell" of a pebble, and so on . . . and perhaps, for termites, the "curved smell" of an arch. I am not inclined to go deeply into this opinion, for the termites' working environment is too confined and too crowded with their colleagues to be saturated with anything but a "termite smell". How could they be able to recognize all the separate scents? Further experiments would certainly be required before coming to any conclusion.

In a word, it is, as Grassé says, work itself that stimulates the worker. It has *stigmeric* properties (from two Greek words meaning "I incite to work"); even if the workers on the shift are continually renewed, it is the *work* which, by its height and shape, *ensures its own organization.* But certain difficulties still remain: for example, if the work is not there, the worker will go to fetch it. Grassé saw two construction gangs some distance apart at two ends of a tunnel, the straightness of which made it clear that the aim was for the two gangs to join up. . . . On another

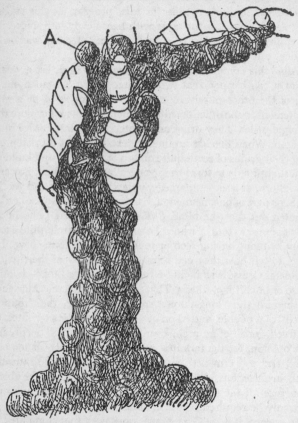

Fig. 32. Termites building an arch (after Grassé). A represents a pellet of excrement used for the construction. Although the workers are blind, the other half of the arch built independently by another team will meet the first one exactly.

occasion Grassé observed at some termitaries in Oubangin that the workers were carrying out their construction with a certain clay that had to be brought from some twelve metres below the nest. For this they had to come and go along a very long and complicated path, and the many obstacles they had to overcome on the way did not deter them in the least. *They did not*

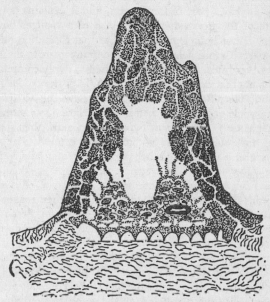

Fig. 33. Nest of the *Bellicositermes natalensis*, after Grassé. The royal chamber is outlined in thick black; *in the centre* are the fungus-beds; *below*, the conical pillars.

wait passively for a stimulus but went out to find it. It is a general characteristic of living organisms that they do not so much give themselves up to an urge as provoke it. *Undoubtedly a task on the way to completion incites the worker to further activity, but, over and above that, the worker seeks work to accomplish.*

How shall we judge the theory of stigmergy? It is difficult at present for we have insufficient data. I think we must accept

Grassé's ideas conceiving the early stages of construction. His theory should inspire further experiments: for instance, the area of building might be determined at will by piling up three or four pellets in one particular place during the stage of unco-ordination.

But, as it often happens in the natural sciences, the application of the theory can be only partial. It gives a satisfactory explanation of the early part of the process, and it explains how the structure of the nest assumes the form of a sponge. But the termitary is not homogeneous and has that formation only in certain parts. Grassé, who has given so many years to the study of African termites, has himself described the very complicated arrangement of a *Bellicositermes* nest with, for instance, the huge, unaccountable pillars at its base. These have the accuracy of machine-made objects, and their size, in relation to a termite, is the same as that of the Great Pyramid to a man; they support nothing since the bases do not touch the ground. The royal chamber also is unlike the other parts of the nest, and so is the exterior cover of the whole structure. It appears, then, that termites react differently according to what part of the nest they are in; perhaps this is due to a change in their reactions or perhaps different building gangs have different "reactional themes", and consequently are not affected in the same way by the work under construction.

Even in nests of very homogeneous structure, like those of bees, certain observations make it difficult to accept the stig-mergy theory completely after the first stage. For example, bees may re-establish the parallelism of the comb, not only by pro-longing the edge of the cells, but also by shifting the foundation if it is found to be too near another comb facing it. And, above all, can you describe as stigmergic a sculptor who carves a statue from a block of stone by *taking material away* from it? For that is what bees do, as Darchen has well shown. If, in a small hive where the impulse to build is not very strong, one sticks fragments of shaped wax in various forms on the ceiling, one soon notices that *the border is cut out till, in some cases, it exactly resembles the ellipsoid form of the original comb* with which the workers always begin. It is as if a termite, faced with a mass of earth, removed (instead of adding) all that was superfluous until the form of a pillar emerged.

Examples like these sometimes make me wonder whether even the theory of stigmergy will allow us to by-pass the strained hypothesis of workers following a preconceived plan; unless, perhaps, the theory of inter-connexion can explain things better.

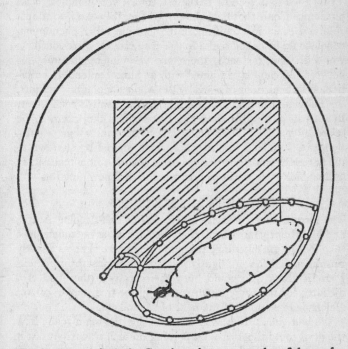

Fig. 34. An experiment by Grassé on the reconstruction of the royal chamber. The termite queen, *right*, put into a vessel with a handful of workers and some building material, is very quickly surrounded by columns (the small circles) which are joined up into a wall, then into a continuous vaulted ceiling.

CHAPTER THREE

FRENZIED JOURNEYS: MIGRATIONS

WHAT WE have seen up to the present are very complex social phenomena, ones that have developed over the years into adaptable and exact mechanisms. But there are other phenomena which could be called social in the crude state, or pre-social, as you wish. For instance, there are very intense interactions between individuals combined with an almost automatic imitation of the behaviour of a few by the whole population; in short, the individual insects are strongly drawn towards each other. But it can go further than that: sometimes they seem to be seized by madness. Their actions, copied by the whole population, can often lead to a needless death, shared by all down to the last individual. The best example of such a fit of madness in animal biology is locust clouds, which I will now describe.

RED CLOUDS THAT HIDE THE SUN

The lorry bumped heavily along, full of half-sleeping Arabs; and I was hardly more lucid than they, for it was not much more than four in the morning, although the eastern sky was already turning pink. We were tightly pressed one against the other and I envied those who could bury their heads in the hood of a djellaba, for the early dawn in the jujube-tree steppes of Ahl Chichaoua has none of the heat of Marrakesh.

We had gone a long way after leaving the main road; how the devil could the lorries keep going on such tracks? At last it was daybreak. I had not known that the jujube trees were red. "Djerad," said a Berber, pointing with his thumb—"locusts!" I had been told about them, but I had never imagined them like this. As far as we could see, and, as we later learned, over a hundred square kilometres, the branches of the bushes were loaded to breaking-point with the big desert locusts, as long and as thick as a finger. They were *Schistocerca gregaria* adults, dark red in colour and rendered immobile by the cool night. A few of the brighter sparks were already busy filling sacks with them: they would sell them that evening, suitably boiled, on the Djmaa el Fna square. The Arabs eat them as we would shrimps; but to a European palate they are horrible, with an after-taste of stale fish soup. I was astonished. To think that I had started to write a thesis on the biology of these creatures, that their migrations had been described to me dozens of times, but that I had never actually seen them in the field! They hardly moved, except when the sun's rays fell directly on them. You could pick them up and examine them at your leisure: they gazed at you with their large eyes striped with reddish brown, probably without seeing you; they hardly even moved their feet. The Arabs started to throw out the large pellets of insecticide as the sun rose. The locusts came slowly down from the trees and began to walk; now if you tried to catch them they jumped away and flew hither and thither. Towards nine o'clock it started to get hot. You could not walk without holding an arm in front of your face; not that the insects try to bite you, for they are in no way aggressive, but each one weighs two or three grams and it is far from pleasant when they hit you in the face at full speed. All of a sudden an enormous shadow covered us. On the horizon, it seemed as if a red cloud had risen. The locusts were flying in millions; and it is quite true, not just a traveller's tale, as I had often thought: *they do hide the sun*. Blinded, stunned, lost in a cloud of locusts, we took refuge in the lorry; yet ten minutes later complete calm reigned again. All that could be seen were a few sick or injured insects whilst, on the horizon, a red-brown cloud, standing out against the blue sky, moved rapidly towards Marrakesh.

You must know that nothing explains the direction they take, or their arrival, or their departure. Naturally the first hypothesis put forwards was the simplest: locusts (and migratory animals in general) leave a place when there is nothing left to eat, to find food elsewhere. *But this is completely wrong (as regards both locusts and other migratory animals)*. Quite the contrary is the case; they may easily leave a good pasture, far from being exhausted, to lose themselves in the desert, or to fall by millions into the sea. Such a thing happened not far from Rabat and the ebb and flow of the tide piled enormous heaps of rotting insects on the beaches and stopped bathing for a week. Such collective suicide is not unparalleled among migratory animals, as we shall soon see.

A few years later I found myself in the middle of another migration, with the Corsican sun roasting my shoulders. This time it was a smaller creature, the Moroccan locust, *Dociostaurus maroccanus*. The invasion was not on such a big scale, and in any case the locusts surrounding me were only in the larval stages and did not yet have any wings. Nevertheless it was most impressive: larvae of all ages tirelessly moved forward like some sort of inexorable machine towards Ajaccio. No doubt they could see me from some way off; behind me they moved a metre and a half from their route, to reform their column a metre or two in front of me. The word "column" is perhaps not quite right; it was more like a broad front, with the density of the creatures not always the same. Once again, as dusk fell and the evening grew cool the indefatigable marchers stopped and started to climb bushes in a more or less dispersed order. But when the morning sun reanimated them with its rays, they came down again and took up the march in *exactly the same direction* as that of the previous day. Nothing stops them: if there is a wall in the way they will go round it or climb over it: but if there is a door left open they will use that, for, like all animals, they obey the law of least effort and do not tire themselves unnecessarily. They throw themselves into water, fill up ditches with their own bodies, extinguish flaming barrages hastily erected to stop them so that the bulk of the swarm can find a path over the roasted vanguard. Life has thus become doubly crazy: first regarding the choice of direction, blind, often leading to mass suicide, as we have just seen, and secondly regarding the proliferation of the swarm. The figures are startling: each of the large desert locusts weighs 2 or 3 grams, as I said; the cloud can cover a hundred square kilometres and the weight of the locusts that make it up must be more than 50,000 tons. When one has seen them, as I have, one can no longer doubt that they can stop a train, for the wheels in crushing them skid on the rails. Not far from Setif they had attacked the roadside poplars, and browsed on the

leaves and young bark; the trees were drying up under the
scorching South Constantine sun; total duration of the operation: about ten minutes. And if you walked under the trees you
were up to your ankles in locust droppings.

THE MECHANISM OF THE ATTACK

A large number of research workers in many countries have
attacked this strange problem, as you may well imagine. It all
started with a happy discovery of Uvarov's about 1925; this
marks the point of separation between what we begin to understand (the mechanism that leads to proliferation) and what we
do not yet understand (what determines the inexorable and
often absurd directions migrating locusts take).

About 1925 Uvarov was studying two closely related species
of locust both belonging to the genus *Locusta*, one green, the
other black and red. On his return from a journey he found
some black locusts in the green locusts' cage. He thought his
laboratory assistant had been careless, but the man stoutly
denied the accusation and declared that he had actually seen
the green locusts become black! Uvarov's great merit in this
case was that he did not angrily sack this joker on the spot;
something in his remarks caused him to reflect for a moment.
Why was he in this receptive state at that moment? What part
had been played by those various small subconscious observations that one hardly knows one has made until they suddenly
emerge in a new context? Why all of a sudden does the world
seem quite new to us laboratory people? And why, under other
circumstances, does it still seem so opaque, though in reality we
have all the facts needed to solve our problem, yet only do so ten
or twenty years later? How much I have wished, with others as
visionary as myself, to construct not the *machina speculatrix* but
the *machina implicatrix* which will weigh all the implications of a
hypothesis and the possible results of an observation without
being obstructed by the unconscious inhibitions of scientists.

Uvarov's hypothesis was not absurd, but one had to be pretty
courageous to propose it: *the green locusts changed colour because
they were in a cloud*, under a mysterious influence exerted by their
companions. Would the black ones then become green if kept in

isolation? Yes. Uvarov was not long in checking this: two species of locusts, believed to be distinct, were really one, and transformed themselves from one to the other according to the conditions under which they lived, isolated or grouped.

But what were the stimuli that passed between individuals and were sufficiently powerful to modify their colouring and, step by step, their whole physiology? This was the subject given to me in 1937 for a thesis, with the request that I sort it out as best I could amidst the prevailing total darkness. I was a bit frightened but at the same time very excited, because the phenomenon was beautiful, undeniable and nobody knew a thing about it: I love that. I have told the story elsewhere of how much effort and how many years it took before a little light was thrown on the subject. It all sprang from an experiment that was very successful: when a young green locust is put into a bottle and the bottle put into the middle of a group of black locusts, the green one becomes black. If the same experiment is done in the dark, the locust stays green. One is thus forced to the conclusion that it is *the sight of its locust companions that brings about the colour change*, no doubt through a series of hormone productions triggered by a visual impulse. This seemed extremely odd at the time, but today a series of studies on parallel subjects have prepared us for the admission of such a hypothesis. Moreover the eyes were not the only organs concerned; the antennae also played a part, as some other experiments were quick to show. Not only does the colour change, but the appetite becomes voracious at the same time and general activity increases considerably.

I noted in addition another still stranger phenomenon, without having time to go into it: a hereditary factor was also involved. My actual subject was the general physiology of the green larvae compared with the black ones, which obliged me to distribute food every morning to 200 cages where the young locusts were becoming more or less green; but I soon found that, to get a good percentage of green locusts, complete isolation was not the only necessary condition. *The conditions under which the mother was raised also had to be considered.* She had to be raised in isolation and herself have developed from a green larva; the male had to have met her only at the time of mating, after which he had to be taken away at once, or the yellow colour would begin to show on his skin, the mark of the gregarious phase, and the female, instead of giving birth to quiet green larvae, would produce a bunch of black "fly" of tremendous activity.

This curious behaviour was studied many years later and developed splendidly. First of all came the work of Ellis (1954) who found that the stimulus inducing the gregarious yellowing of the males was chemical in nature; it is an emanation which is effective only at a short distance and acts upon a sense intermediate between taste and smell; the antennae are the sentient organs and I had often noted that if a number of locusts with the antennae cut off are grouped, yellowing does not take place, or only very slightly. However, a still more curious fact was brought to light (and similar things are found with birds and rats): *the development of the ovaries of the females is stimulated by the presence of males even in the absence of any mating.* This is reciprocal, moreover, and if the females are numerous they also stimulate the males to a more rapid sexual development and consequent yellowing. But what I never saw, and Ellis clearly did, was that *young* males or females not only do not accelerate by their presence the yellowing of a nearly mature male, but slow it down a little. Finally and above all, as Albrecht showed, grouping of the females has an enormous influence on their fecundity. The weight of the young at birth depends on the "phase" of the mother and even of the grandmother; it is the same with the number of their oviaroles. Gregarious females, contrary to what was thought, give rise to *fewer young*, because they have fewer oviaroles, than the *solitaria*, but the young are bigger and come from eggs richer in yolk than those of female solitaria. The parents' influence is so powerful that at least three successive generations are needed to produce perfected gregaria from solitaria and vice versa (starting obviously from two diametrically opposed points). Albrecht's discoveries, based on an incredible number of painstaking observations done with great care, will no doubt allow us in the near future to distinguish in the field the young *congregans* from the young *dissocians* stages; that is to say those coming from the solitary phase and tending to the gregarious and vice versa. This is of the utmost importance in the fight against locusts, which needs accurate forecasts on the future state of locusts found in the field.

The Mechanism of swarm formation. This long detour has been necessary in order to return to one starting-point: the explanation of the locust swarms. First of all we must note that these creatures have a permanent habitat in certain areas, called "gregarious areas", far removed from their normal invasion zone. For example the vast outbreak area of the locust (*Schisto-cerca gregaria*) that devastates Algeria covers the southern border

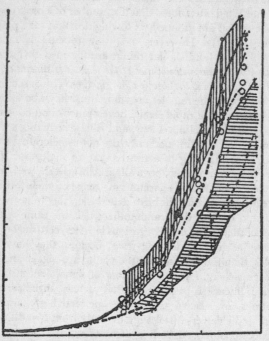

Fig. 35. Action of the group on growth (group effect) in the locust *Zonocerus* (after Vuillaume).
Abcissae, time in days. *Ordinates,* weight. The grouped locusts (vertical shading) grow more quickly than the solitary locusts (horizontal shading).

of the Sahara. Here the solitaria types live, if one may so put it, on the edge of extinction. The solitaria may well lay more eggs than the gregaria, as Albrecht has shown, but this does not help them very much, as the pitiless climate of the great desert kills almost all the offspring. It appears that every now and then the stranglehold of the climate slackens; rather more of the young reach maturity, and if their young and the young of their young also live under these more favourable conditions, then the inexorable law of geometrical progression starts to work and the desert groans under its load of locusts. However another phenomenon, a well-established one, comes into play at this moment: as Kennedy saw in the case of *Schistocerca gregaria*, and as Vuillaume studied in detail in the case of *Zonocerus* (an East African locust), these insects have good sight and are drawn a considerable distance by any salient object, such as a tree or bush. When they are spread over a large area there is a good chance that there will be trees in the neighbourhood; the locusts climb them, and since they are living close to each other, the stage of turning to gregaria is started: their colour gets darker, their activity increases enormously, and soon immense troops of hoppers deludedly march forward, always in the same direction, a direction renewed every day after the night's repose. What are the factors which enable them to maintain this direction?

MAINTENANCE OF DIRECTION

I must say that up to the present we know very little. A lot of hypotheses have been put forward, and I think I have been guilty of two or three supplementary ones, worth no more than the others. The wind has been thought of: but near the ground the wind is so much contorted by the lie of the land that it is turned into a series of whirlwinds that are useless as direction indicators; and if Haskell has just proved that young locusts move up wind, one does not see how this laboratory experiment can be related to field experience. We then thought of the sun. I used to oppose this "sun hypothesis", the pet theory of all English writers, my objection being that the sun moves but the locusts keep on in the same direction. Then came the discovery that bees know how to compensate very precisely for the sun's displacement, and follow a constant course to get back to the nest, and this ability has now been found in a whole crowd of other insects. This is true, but the march of the locust hoppers cannot be compared to the bees' return to the nest: they just go straight forward, through fire and water if these stand in the way. And why the devil do they take up the same direction again in the mornings, since they have been sheltering in the bushes and their bodies, lost in the branches, have taken up all sorts of positions so that it is quite impossible for them to retain any memory of direction? I battered my head for a long time against this wall; it is the same maddening problem of the "hypnotic" adhesion to a route found with all migrating animals.

Is there any way out of this difficulty? Yes; the English have given it the word *"imprinting"* and the Germans *"prägung"*. Lorenz and his pupils studied it in young birds for a long time. When a gosling hatches in an incubator, and has never seen a creature of its own species, it will follow the first moving object that catches its attention, be it a man or a cushion dragged along by a piece of string, it does not matter which. This special sensitivity at the start of life only lasts a few hours and is completely lost later on; but the creature will be marked by this "imprinting" all its life. For example a young gosling or jackdaw which has been imprinted in this way with a man will consider him its own species, and their true fellows will be strangers to them, sometimes so definitely that they have the greatest difficulty in reproducing themselves, and in certain cases cannot do so at all. Perhaps, then, the young locusts, at hatching, fix themselves on a certain position of the sun, that they never forget, because they first saw it there during a particularly sensitive period of their lives, no doubt a very short period. They could then follow it all their lives, compensating for the daily movement of the orb as do many other insects.

LEMMINGS AND THE MIGRATION OF MAMMALS

According to many authors this is another case of a typical biological frenzy. That good old chronicler Olaus Magnus had already written about it in the sixteenth century, dumbfounded to see these little rodents, usually so timid and solitary, proliferate and migrate by millions. No doubt some devil from hell was responsible for it and the lemmings should have been exorcized, as at one time were caterpillars and grasshoppers. The lemming lives in the Scandinavian countries and is hardly ever seen, for it is both timid and nocturnal. But every three or four years it seems to go mad. It starts to proliferate, is seen in broad daylight, outruns the limits of its habitat and starts to migrate. During the migrations lemmings keep themselves out of contact with each other, unlike locusts; they all go in the same direction and do not get together in really big masses except when faced by some considerable obstacle, such as a river; then they throw themselves into the water by millions, swim straight forward and scale all obstacles in their path, such as a boat, which can become so loaded with them that it sinks (a lemming is hardly bigger than a mouse). They will venture out to sea, says Laurent, and in 1868 "a steamship going at full speed in Trondjheim fjord had to pass through a veritable soup of lemmings swimming away in the waves with their innumerable little heads puncturing the surface as far as the eye could see". At these times the lemming, usually so timid, is not afraid to be seen in the middle of towns, even in the houses of men, whom it will even sometimes attack. Each individual displays a manic imitation of its fellows; if one lemming throws itself into a ditch, millions will follow suit until it is full to overflowing. They may all throw themselves into space from a viaduct, or they will swim straight out to sea until they are all drowned. In short, they seem to go mad.

I must, however, draw attention to a paper which has just appeared in the old and celebrated German periodical *Zeitschrift für Morphologie und Ökologie der Tiere*. There Dr Frank discusses the problem of the lemmings from first principles and undertakes a complete revision of their biology. What struck him more than anything else was their *antisocial* tendencies: each individual lives far away from the others and the female cannot wait to get rid of her offspring as soon as they can fend for themselves. The male very timidly approaches the female's lair for the purposes of mating, after which he is thrown out without ceremony. Thus, says Dr Frank, it is scarcely possible that lemmings get together in huge migrating crowds, as they are said to do; such a strongly developed antisocial instinct would never permit such a thing to occur. Moreover no "serious author" has ever described a lemming migration, only writers with no scientific training have done so. I am not, myself, a lemming specialist and can only give an opinion on general principles. It seems to me that here we have an example of the two tendencies that animate scientists, and enable science to advance: the minimizing tendency, characteristic of minds more critical than intuitive, as opposed to the intuitive, which often supposes more than it deduces. Frank had never seen a big lemming migration. He then, perhaps rather hastily, jumped to the conclusion that they did not exist and that authors who wrote of them were not worthy of credit. . . . Might one remind him that the antisocial tendencies of lemmings are an incontestably true fact, but no doubt only in the non-swarming periods? Locusts are also solitary to start with and Ellis has shown that they avoid one another as soon as the antennae touch the body of another locust; but God knows they become sufficiently gregarious at times!

On the other hand we can find many examples of mammals who migrate whilst appearing to be "out of their minds", just like the lemming: the grey American squirrel, for example, migrates in hordes that can contain several hundreds of millions of individuals. There is also the springbok antelope of South Africa: masses of them migrate in such dense crowds that lions can be trapped in their midst, unable to escape in spite of all the effort and fury that can be imagined. Frank said that the migration of lemmings was no doubt due to the impoverishment of the northern vegetation, which every now and then forces the animals to seek their food elsewhere. I would like to think so, but I have my doubts, because we have so many cases of migration where there is no lack of food, and not only in the case of locusts. Springboks, for instance, will leave excellent pasture for arid zones where they die of hunger; or they will throw themselves by millions into the sea. It must be said that the springbok usually has a gregarious tendency, and that it will even join up with individuals of other species, such as ostriches.

It seems to me that this madness which seizes upon certain migrating mammals is the sign of a deep neuro-endocrine unbalance, with no direct and exact relationship to food supply; but perhaps related to some meteorological factor that has been overlooked. Some workers have spoken of sun spots, but one needs to look beyond that. Perhaps so strange a phenomenon, a frenzy so contrary to the conservation of the species, demands an explanation more complex than those we usually discuss.

Fig. 36. The Lemming (*Lemmus lemmus*) from the *Traité de zoologie*, Grassé.

MICE AGAINST MALTHUS

As the temple of the sciences grows, it becomes as complicated as the Babylonian library so dear to Borges. Not only can specialists in one discipline no longer talk to scientists in another, but even within the same discipline there is difficulty, though the respective fields of research may differ but little. It also often happens that lectures and meetings take place where most interesting problems are discussed, and one only hears about them much too late, even though one really ought to have been there. When are we going to reform our archaic publication system and feed punched cards into a big machine? As a celebrated English nuclear physicist said, "It might not be much better, but in any case it could not be worse."

When Coon, Christian, Snyder and Ratcliff thought it a good idea to get together under the auspices of the American Academy of Science I asked myself if they really knew what they were doing. It seems, however, that the chairman, C. S. Coon, did realize the implications when he said, "Unless I am hopelessly naïve, it seems to me that Christian and his colleagues have opened a new door in the study of evolution, a non-Malthusian natural selection."

This is such an extreme way of talking, especially in an Anglo-Saxon country, that it needs looking at more closely. What, then, is so remarkable about Christian's research? It is easily explained in a few words: you remember the Malthusian theory, which has influenced not only biology but philosophy as well: consumers of food increase more rapidly than does the production of food, with the result that the latter regulates the former. Consequently human reproduction should be planned since human population increase has no other regulator than famine. Of course it is the same with animals. Now, according to Christian, all this is untrue: *there is an automatic population regulating mechanism at work.* It has been found in all animals where efforts have been made to find it. It is exact and quite independent of the food available. This is altogether revolutionary, though in my opinion perfectly well proved by a mass of work done in two or three American laboratories, and very little known in Europe.

To tell the truth, all this actually started a long time ago; as is always the case in science, there were pioneers in the field. Crew and Moïskaïa in *Biologia generalis* in 1931, Vetulani in the same journal the same year, then Retzlaff, again in that journal, in May 1937. All these workers noted that grouped mice and isolated mice did not behave in the same fashion and that the physiology of the two was greatly different. These authors worked in close contact with each other, and one feels that a school of experimental animal sociology was waiting to be born. But the time was not ripe, as I discovered myself when, a few years later, I found yet stranger things happening among migrating locusts. The idea that individuals of a species can act as a special stimulus to each other was something that had. certainly not penetrated scientific heads in 1939. Here once again was the phenomenon of the "allergy to innovation" always present in science, which does so much to prevent progress.

In any case, much later, Christian and his colleagues found that a couple of mice, male and female, cannot normally reproduce themselves in a cage. Several couples must be present; otherwise a number of females will not give birth, while others start a gestation, but the foetus is quickly reabsorbed. The presence of the male, moreover, is vital for the successful working of the female reproductive system, even if the mice are stopped from mating by a wire screen. It is not enough for him to do his job as a male and then be taken out, the females must be subjected to the continual excitement of his presence, and perhaps of his smell.

Moreover these were but very generalized observations; in our zootechnology laboratory at Jouy, the most modern in Europe, a group of workers showed clearly that the sow's gonads do not develop normally if she is prevented from hearing and smelling the boar. However, when a much more numerous group of mice is put in a cage another set of phenomena at once occurs: reproduction is normal when the population density is low enough, but if you allow reproduction to go on at will, supplying the animals with plenty of food and water, it will be seen that the mortality of the young increases as density increases, until a point is reached where reproduction stops altogether. At the same time the adrenal glands grow and show signs of greatly increased activity.

If some of the population is removed, reproduction starts again, and the size of the adrenal glands is reduced. This variation in the size of the adrenal glands, and a few other things that I cannot particularize here, clearly shows that a regulating mechanism is at work which, through a whole series of interacting hormones, brings about a progressive reduction of the birth-rate and can even reach the point of complete suppression of births. This is the typical "anti-Malthusian" phenomenon.

The surprised reader obviously will not fail to put forward some objections to this view. Everyone knows, he might well say, that if you shut up an excessive number of animals in a cage a whole series of physiological and pathological troubles will follow. Nothing very mysterious about that. Our reply is that the word "overpopulation" is relative here, and that the blockage of reproduction occurs long before the breeders consider the animals to be too crowded; moreover, if in place of the white mouse, a gentle and easily managed creature, you use the fierce grey mouse, then a very modest population density is enough to bring about the blockage. But, you might still object, what is there to tell us that it is density of population that is responsible for all this? The American authors asked themselves the same question, and they moved populations being subjected to "group sterilization" into very large terrains where apparently they should have felt at their ease. Now a mysterious effect was produced; the "living space" effect does not work alone, and the blockage phenomena continued as before, no doubt because the rodents continued to crowd together and the living space then only had a relative importance: in short there is a certain average density, very different for each species, above which the mysterious regulator of the adrenal glands and the pituitary gland starts inexorably first to reduce, and then to suppress, reproduction.

Looked at in this way the phenomenon is truly strange, but still relatively simple to interpret. Unfortunately we at once have to complicate it, because we must take note of the hierarchy, or "pecking order". A crowd of rats or mice is not just an unorganized magma of flesh, as the layman would like to think. On the contrary an *alpha* animal can be found, which one might call the chief (but isn't that too anthropomorphic a way of talking?); he may chastise everybody, he eats before all the others, he monopolizes the dominant females (for there is a parallel female hierarchy too, a quite distinct one), and he stops everyone copulating, except when he is asleep. Beneath him we find *beta*, who only gets the chief's left-overs, but freely takes it out of the others beneath him. And so on, using, if need be, all the letters of the Greek alphabet, until we get to omega, who cannot mate, is battered day and night and only eats by stealth. He often dies from sheer physiological misery, always provided the others do not kill him first. Now it has been noted that the increase in the population of rodents is less rapid if part of the population is frequently changed; for example, if you take out 15 per cent of a population in its increasing phase and replace it with the same number of stranger rats, then the increase in numbers stops. This must be due to the fact that social relationships of dominance and subordination are greatly upset, and can only be re-established after an interval. Statistical studies have shown that nearly all the young come from the dominant individuals. Then, you could say, it is quite possible that the modifications to the adrenal glands of which you have spoken are strictly related to the continual quarrels taking place in defence of the ranking order. Not at all! In practice these quarrels are not so very frequent; dominance is quickly established, and thereafter the least pretence of attack put on by *alpha* serves to drive away *gamma* or *delta*, who will not be so daring again. Moreover rats have been put in situations where quarrels were frequent, and attempts have been made, with no success, to find a correlation between the number and seriousness of their wounds and the changes in the weight of the adrenal glands. All this must be more complicated than our rudimentary theories allow us to see.

Of course all these experiments have been done in laboratories. What indication is there that the same results would be obtained in the field? Well, helped by the laboratory work, scientists have found exact parallels in the open field. For example, we have spoken of the wild hordes of lemmings which increase to a fantastic extent, then blindly migrate and sometimes throw themselves *en masse* into the sea. For a long time this went completely unexplained until one day, after Christian had announced his work, someone had the idea of looking at the adrenal glands: these were, as might be expected, very big and heavy, which to a great extent explained the creatures' abnormal excitement and behaviour. It was once again a question of reaction to crowding. In the case of voles, as far

as one can find out by trapping, the maturation of the young
is arrested as soon as the population density increases a little;
so that during the summer one sees an increasing number of
immature but quite old individuals; moreover the growth of
the males is more affected than that of the females. Kalela has
shown that food and climate are of no importance in this matter.
In the case of rats in towns, trapping by the official rat catchers
allows statistics on the subject to be obtained and the weight of
the adrenal glands always drops after a trapping campaign has
reduced the population.

The Sika deer (*Cervus nippon*) which inhabit, among other
places, the grounds of the President of the Republic at Ram-
bouillet, very quickly experience a high mortality the moment
their population density rises above 1 per 40 acres; at the same
time the weight of the adrenal glands increases, to decrease
again when the population falls below a certain level. Just
before the phase of mortality begins, the growth of the young
drops by 40 per cent. Hepatitic lesions appear, or glomeruli,
signs of a lowering of resistance to normal diseases, which the
deer's body cannot now withstand. The hyperactivity of the
adrenal glands may well be responsible for this lowered resistance, since massive injections of cortisone, a highly typical
adrenaline substance, have precisely the effect of lowering (as
all doctors know) the ability to resist disease.

Obviously a last question springs to mind: if this is the case
with animals, what happens with man? Well, my reply may
perhaps surprise the reader: it is possible (for we still really
know nothing) that all this does not apply to man. With man it
is another story, for the group effects we know of in insects take
on quite another aspect when we come to the higher social
animals. With insects the considerable population density
(that of the hive or ant hill, for example) has no bad effect,
on the contrary, however enormous the population may be in a
restricted space. Now man is the sole mammal leading a really
social life, comparable to that of the ants. Other mammals are
but sub-social. And up to the present the little that we know
about the development of children in towns suggests that they
increase more rapidly there than in the country. So, after all,

it may well be that men are Malthusian and mice anti-Malthusian, but what is most certain is that we know nothing; suddenly a vast field is opening before our eyes, and it is up to us to explore it.

BIRD MIGRATION

I will write almost nothing of this, because it is a phenomenon of another kind, though none the less strange for all that; birds migrate in an ordered, I might almost say a methodical, manner, which periodically takes them to warmer or cooler regions, where living conditions will be much better. There is nothing here comparable to the folly of locusts and lemmings.

A spectacular phenomenon precedes the migration of birds; it is the assembly, side by side, sometimes in great numbers, of individuals who are habitually shy and who live in isolation. Swallows are an example known to everyone. Sometimes other species, even non-migratory ones, are caught up in the assembly. The length of the journey is often unbelievable. For example, the Pacific islands are inhabited during the northern winter by birds from Siberia or Alaska! One species (*Arenaria interpres*) nests in the arctic region and spends the winter in New Zealand and Chile. But the long-distance champion is probably the arctic tern (*Sterna paradisea*) which nests in the arctic up to latitude 75° but passes the winter on the temperate shores of the southern hemisphere and has been found at times in the antarctic! Its double journey must be between 12,000 and 15,000 kilometres.

PART TWO

A RETURN TO OUR WORLD:
SOCIETIES OF HIGHER ANIMALS

CHAPTER FOUR

EXAMINATION OF THE ELEMENTS OF SOCIAL LIFE

As I said at the beginning we are now going to enter another world, one much closer to us than that of the insects.

Solitary and gregarious life. Social relationships can only exist where several animals are in close proximity; that is to say, they must not have too marked a preference for a solitary life. Not all animals are sociable. Some, such as the lemming, discussed in the previous chapter, can hardly bear the presence of their mate during the time needed for copulation and turn their partner away almost immediately afterwards. At one time it was believed that species leading solitary lives had a bigger struggle to find food than those leading gregarious ones and that this led the individuals of certain species to keep their distance from each other. But the reality is more complex: for example, some herbivores are shy and solitary, and yet the savannah or prairie provides them with plenty of food. Others live on this same prairie in immense herds. It must, then, be a question of utterly different behaviour patterns. The life of the solitary species is linked in every case to the possession or very strict demarcation of a *territory*. This idea, relatively new in animal psychology, has shown itself to be basic.

Territory. A territory is a more or less extended zone, in general bearing a relationship to the size or method or life of the creature: the territory of big hunting mammals is relatively enormous. Its boundaries are very well known by the legitimate owner, and are marked in the case of mammals with the excretions from certain scent glands, which are often strangely placed: certain antelopes, for example, mark twigs on their boundaries with the secretion from a periorbital gland. Bears rub their backs against trees and stones and leave greasy marks. If dogs raise their legs so often, it is to assure themselves of their rights in respect of certain trees, stones or even a car parked in a

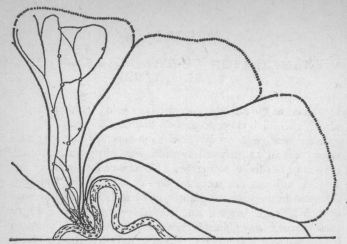

Fig. 37. Plan of some hippopotamus territories on the borders of the Rat-shura, Congo. They are all pear-shaped. The small circles indicate defecation spots, the fine lines their routes and x's their shelters. (After Grassé)

yard under their guard; the few drops of urine tell a potential rival that he had better keep his distance. The proprietor, as we shall see, is dominant in his territory, but outside he runs the risk of being attacked.

The same thing is found in the case of fish. Young males move around, generally avoiding the females. But when the mating season arrives they start to think of acquiring a territory. One of the males leaves the group and takes possession of the largest territory possible, all the aquarium if he can. Later a second male, and then a third, also establish themselves, managing to become accepted by the first occupant after· numerous fights, they all try surreptitiously to enlarge their territory at the expense of the first arrival. After a bit all the fish manage to appropriate a patch of the bottom, but the size of the different pieces is naturally near the minimum. This only goes for the species that nest near the bottom, and who thus protect and defend the base of the column of water rising from their territory; but there are other species, such as *Labyrinthici*,

Fig. 38. An antelope marks a twig in its territory by leaving a secretion from the periorbital gland. (After Hediger in Bourlière)

that lay their eggs in a nest of floating scum, and such species are only interested in this area and ignore the bottom. In general the territories of fish are established only during the period of reproduction and the eggs generally remain there until they are hatched. However in the case of the *Cichlidae* the female carries the eggs right out of the male's territory in her mouth, as soon as they are fertilized.

HOW TO RECOGNIZE A RIVAL

This obliges us to discuss Lorenz and Tinbergen's famous theory of releasers, and to do that we must to some extent go into the history of the matter. About the start of our present century animal psychology was being built up on a very narrow basis, in reaction against naïve anthropomorphism, the enemy of science. That was the heyday of *tropism*, a theory put forward by Loeb, his pupils and emulators. The model was the butterfly which will throw itself into a flame, foolish, blind to the danger it runs, as though spitted by the rays of light. Loeb maintained that all animal behaviour was ruled by simple stimuli, which called forth automatic and unadapted reactions. This is quite true, *but only under certain conditions, namely in the laboratories,* where specialists in tropism did all their work.

Fig. 39. *Betta splendens.*
Below: A characteristic releaser appearance.
Above: Normal appearance. (After Hess)

The leaders of the "objectivist" school of ethology were Lorenz and Tinbergen and they were field men, not interested only in insects and lower animals, like Loeb, but also in fish and above all in birds. They soon saw that in the fully natural state animal reactions appear to be very different. It is never a single isolated stimulus, such as a ray of light, that sets off a reaction, but special "objects", details of colour and body shape, in short *releasers.* The best way of understanding the place of releasers, of which we shall find a great quantity, is to quote the classic example, the stickleback at breeding time.

The red belly of the stickleback as a releaser. Before the breeding season starts the sticklebacks live in a shoal, but soon the males separate out and, as we have seen, start looking for territories. At the same time their eyes take on a brilliant blue colour, the back turns from brown to green and the belly becomes red. As soon as a male clothed in these colours gets into another's territory the owner attacks, although usually the two rivals

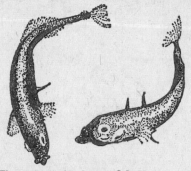

Fig. 40. Preliminaries to a fight between male sticklebacks. The one on the left, at the edge of its territory, is threatening the one on the right. (After Tinbergen)

do not fight to the bitter end and in fact in most cases all the owner of the territory has to do is to "threaten" the intruder. The threatening pose is a special one: the spines of the dorsal fin are raised, the mouth is opened as if to bite, the body is turned head downwards and the tail lashed as if the fish were trying to bury itself in the sand. Sometimes the spine of the ventral fin is raised as well.

Generally this is enough and the intruder leaves. Now it is not just the colour red, nor a red spot, that releases the proprietor's anger: *it is any oblong object with a red underside*, even if to the human eye it in no way resembles a stickleback. Any piece of plastic more or less oblong in shape, or even spherical, will bring about the release, as long as its underside is red: it will be furiously attacked. *This oblong object with a red underside* is the *releaser* that leads to an attack. The effectiveness of such releaser objects depends on a balance of a number of stimuli which the object contains: for instance a piece of plastic which only vaguely resembles a fish, but which has a red underside, gives as great a reaction as a very exact copy of a fish having a pale pink

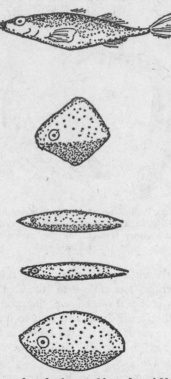

Fig. 41. Examples of releaser objects for sticklebacks. The uppermost one (a fish with a light ventral surface) is not effective. The remaining ones are cruder, but all have the underside red and are releasers. (After Tinbergen)

ventral surface, or a red colour on only a very small area. *It is the total make-up of characteristics that counts,* and the partial lack of one can be made up for by an abundance of another: moreover we should note that in the case of birds, and sometimes also fish, particularly during the courting ceremonies, colours or special bodily characteristics that differentiate the sexes are often insistently presented to the partner in the course of a veritable dance: the English have called this the "display": the creature displays all its sexual characteristics in order to make them easily visible to its partner.

But let us get back to the stickleback. The male, freed for a little while from the attentions of his rivals, starts to build a nest. As soon as it is finished his nuptial colours become yet more intense and he actively parades in his territory and to some extent outside it. All this time the females, who are not in the least interested in nest building, move around in a shoal and some reach sexual maturity; then the ventral surface turns a brilliant silver colour and fills with fully developed eggs. The males carry out a sort of dance in front of the females showing their colours; the dance is a series of incomplete loops in which the male suddenly feigns flight from the female and then rapidly dashes back with his mouth open. What is it that causes him to exert himself in this way? It is no longer a long object with a red underside, but *any long object with a swollen underside* like the belly of a female, even if the object in question only faintly resembles a fish. The long object with a swollen underside is the releaser for the nuptial display.

The majority of the females are frightened by the males' display and swim away, but those whose bellies are swollen are not so timid; and there may well be one who, far from fleeing, turns towards the male, pointing her head upwards, which results in the display of her swollen belly. The male swims around her, then moves towards the nest, followed by the female; he pushes his mouth towards the entrance, as if showing it to her. She finally enters, her head protruding on one side and her tail on the other. The male then nuzzles the base of her tail; after a moment or two the female lifts her tail, starts to lay the eggs and then leaves the nest. What first allowed the female to recognize the male was not only his red belly but also his brilliant blue eye, another integral part of the nuptial dress. But at the egg-laying stage these visual stimuli of the male are of no use at all: if the male, or one of the models mentioned above, is shown to the female at this stage, egg-laying will not result: what is necessary—all that is necessary—is to rub the base of the tail, and a glass rod operated by the experimenter can perfectly well replace the male stickleback at this stage.

PERSUASION

We shall see later that most males who carve out and defend a territory always try to fight any rivals attempting to come into it. It also appears that females fear to enter any territory which is not their own. Moreover a female seems to avoid bodily contacts, as do most animals. The male, then, to some extent, has to "persuade" her, to help her overcome her repugnance to such contact. No doubt this is the reason for all sorts of strange rituals, especially among birds. The different "dress" of the two sexes helps to avoid attacks: for the females just do not carry the releasers that would lead to it. However in some species sexes look alike; in such a case the male certainly makes as if to attack, but it is the female who then adopts a special course of action which prevents things getting worse: in effect she avoids making a "male response" to the attack by taking up a characteristic female posture, which is often very like a juvenile posture: in the case of gulls, for instance, the females crouch down and cry, in the position of nestlings asking for food. Moreover, again in the case of gulls, the search for food becomes a positive mania, as we shall see below: the female importunes the male continuously, even if she has just finished eating and even if the male, who has not left her, could not have gone after fish. In the persuasion ritual we also find the male's curious habit of offering gifts to the object of his desire, which I shall describe below.

PARENTS AND OFFSPRING

Obviously very close relationships, "social" or "familial" as you may prefer to call them, are established between parents and their offspring. Various kinds of releasers are found here. For example, when young gull nestlings are ready to ask for food, a few hours after hatching, the parents disgorge a half-digested portion and offer it at the end of a beak, helping the youngsters to take it, at first hesitantly and then with more assurance. This reaction is innate, and the young gull, without having to learn how, looks for its parent's beak. Tinbergen and Baerends found the releasers that guide the young gulls' beaks; in fact you can bring about this characteristic action by showing the very young creature a model of its parent's head, even one roughly cut out of cardboard. All that is needed is that the "beak" of the cardboard head be yellow and carry at its end a clearly visible red spot. If the spot is of another colour the responses will be more numerous than in the case of a beak without a spot, but they will not be as numerous as with the typical red-spotted beak. Baerends and Tinbergen have spent many hours on the Dutch flats, battered by the winds, trying out a whole collection of cardboard heads, with different coloured spots, on young gulls. They found that *everything depends on the contrast* between the colour of the spot and the colour of the background. Strangely enough, on the other hand, the colour of the beak itself is scarcely of importance, except that a completely red beak is "superoptimal": it induced twice as much response as the other colours, including yellow, which is, after all, the natural colour of the mother's beak. By contrast, neither the colour of the head—white, green or black—nor its shape count for anything. Only the beak is important: nevertheless the nestlings see their parents, and even peck their red eyelids with their beaks; but when they are hungry nothing matters except a certain thin, long object with a red spot at the end, that comes very near to them.

An interesting observation on the heron *Nycticorax* was made by Lorenz; the upper part of this bird's head is blue-black with a triple aigrette of three white feathers. When it came near the nest the bird always bowed its head, so that from the nest all that was seen was the top of the head and the white aigrette of feathers. It was then readily accepted. On one occasion when Lorenz had climbed the tree to get a better view, the heron, disturbed by his presence, forgot its bow and was at once attacked by its own young. These recognize their parents because they approach the nest with a special posture or make

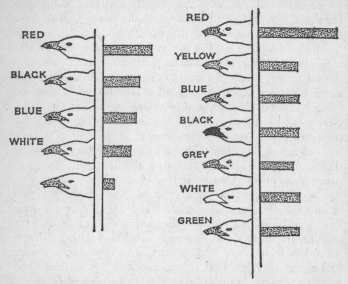

Fig. 42. Tinbergen's outlines showing the reactions to different gull beak forms by the young. The length of the horizontal dotted bands is proportional to the intensity of the reaction.

Left. Effect of the colour of the spot on the end of a uniformly coloured beak.

Right. Effect of beak colour itself. It will be noted that the red beak produces a superoptimal reaction, even though the gull's beak is naturally yellow.

particular movements, and are the only ones of that species to do so.

We have some idea of how the young recognize their parents. In fish, the fry of the *Cichlidae* family, particularly studied by the ethologists, look for the parent on guard and follow that fish everywhere, even if separated from it by a sheet of glass. An anaethetized fish does not attract them, unless it is made to

Fig. 43. The heron *Nycticorax* bows its head when it approaches the nest and thus shows its triple aigrette white feathers; the young recognize it by this means. (After Tinbergen)

move slowly; they flee if the movement becomes too rapid. Now in fact the guardian parent fish always swims slowly, whilst the other is much more lively. This, then, is *the only characteristic that matters*, and not so much the shape or any detail of colouring: the fry will quite happily follow a cardboard disc. The distance the fry leave between themselves and the model increases with the size of the model; they try to keep it the same size in relation to their visual field.

IMPRINTING (PRÄGUNG)

This is a curious phenomenon, discovered by Heinroth and Lorenz, which seems to be of immense importance in the life of the young. *The first moving object seen by the young in the few hours following birth* often leaves an irremovable impression. Heinroth first noticed it in the case of some incubator-hatched goslings. When he took them near a goose they refused to follow her, but on the other hand they would not move an inch away from Heinroth himself. Many other similar cases have been discovered by Lorenz and his school in a great many different animals. It is the first moving object seen that counts, once again, and little does it matter whether it be a man, a creature of the same species, or an object. Lorenz and Tinbergen, for example, once saw an Egyptian goose imprinted with a cushion they were moving in front of it.

The strangest of moving objects will prevail over a stuffed, but immobile, member of their own species. In the paradoxical imprinting on a human being Lorenz nevertheless did find a trace of the innate: his goslings followed him so as always to see him at the same angle; a man is much larger than a goose, thus the goslings followed him at a much greater distance than they would have followed their mother. When Lorenz went into the water, his inseparable goslings imitated him and drew nearer to him the more he immersed himself, because his apparent size was reduced. Finally, when only his head was above water, his companions came and perched on it.

These fixations can last a long time, possibly a lifetime (nevertheless some recent experiments made by the school of Lorenz have shown that they are to some extent reversible). The animal can then become abnormal and incapable of interest in its own species; it would like to have a man as a sexual partner, will court him and bring him food. A chough, raised by Lorenz, used to bring him meal worms; as the professor showed some repugnance to such food, as can well be imagined, the bird tried to put them in his nose or ears. Some of the drawbacks attached to the work of a biologist!

It is probably imprinting that explains the sad story of the "placer sheep". When a ewe followed by an incompletely weaned lamb gets separated from the flock and suddenly dies, the lamb remains attached to some reference point not far from its mother's body, for example a rock or a tree trunk. It will refuse to leave the place, or will try desperately to go back there if carried away, even if the body is decomposing. Later the lamb will refuse to join a flock and will not mate. Sheep farmers in New Zealand well know these "placer sheep" and prefer to slaughter them, as they will never be able to lead a normal life.

HOW DO THE YOUNG RECOGNIZE EACH OTHER?

We hardly know in most cases; it seems that they do not always recognize each other, at any rate as individuals and at the start of life. The only observations known are some interesting ones on the *Cichlidae* fish, whose young live in a shoal. If a glass flask in which some fry have been placed is put into the middle of such a shoal, they all gather round it, and the more fish there are in the flask the more eagerly do they gather. At the beginning of their life, the young *Cichlidae* will even follow an artificial shoal made of drops of wax joined by a metal wire; the colour of the wax does not much matter, even with *Hemichromis*, which will only recognize its parent if the latter bears red spots on its body, yet finds that colour of no importance when it rejoins its brothers and sisters and is no longer with its parents.

HOW DO PARENTS RECOGNIZE THEIR YOUNG?

All stockmen know that after a few days parents recognize their own offspring and kill any little strangers one may try to introduce to them. But such behaviour leads to extraordinary complications in a mouse-raising establishment. A strange mousling can only be introduced to a mother in the three or four days following birth: if this is done a little later, it will be put to death. On the other hand, when two or three females are raised in the same cage, they usually group their young into a single nest and each female gives suck to all of them in turn: I do not know how they do it, for on lifting up a mother mouse, three or four young may often be found hanging from each teat; yet the growth of the young seems to be perfectly normal under these conditions. We discussed the "placer sheep" above; no doubt you know that when a sheep dies the others will refuse to suckle her lamb. Nevertheless, to a human eye, one mousling, or one lamb, looks very like another. Is it, like recognition of the mate (see below), a question of distinguishing minimal differences in facial expressions?

With the *Cichlidae* the adults readily devour the young of other fish species of the same size as their own young; but somehow, we do not know how, they distinguish and spare their own. Noble changed the eggs of a young couple breeding for the first time for those of a related species; the young hatched and were raised, but as soon as the foster parents met young of their own species they ate them. This aberrant behaviour was fixed for good, and in fact stopped them raising young of their own, as they ate them as soon as they hatched. Some form of imprinting must take place during a fairly short, sensitive period at the moment of the first hatching that the young adult sees.

THE PECKING ORDER, OR HIERARCHY

We have already discussed this, and shall do so again, particularly with regard to birds and mammals. You must know, to begin with, that this is a widespread phenomenon: when you see a herd of animals on the move, apparently in no particular order, do not be misled by appearances: the order, on the contrary, is fixed and rigid and there is no place for individual ideas. The Norwegian scientist Schelderup-Ebbe first noticed this among chickens, by counting the numerous pecks they distribute so liberally, but by no means at random. As with mice (see above) there is an *omega* animal who receives beatings from everyone, at times so severe that he is killed; at the top there is an *alpha* creature who beats everybody and is beaten by no one; between these two extremes there is every intermediate stage. And the same thing is found in regard to mating.

With male mice, the only way of moderating the aggressive nature of these ferocious animals is by letting them out into unknown surroundings. The male starts by exploring the place minutely and cautiously; after a longish time, which may vary, he stops keeping to the walls and risks going into the centre. After a few days the mouse plucks up his courage and with no hesitation enters a box, or some shelter he has chosen. When two mice meet they both hastily retire. But at the next meeting there will be some slight difference in the bearing of one of them, who does not draw back. He may even threaten an attack when the other approaches. If one has had time to explore the territory, to appropriate it, he manifests much more assurance and hostility at the introduction of another mouse. The males continue their persecution of the under animals, whereas females quickly give it up. When the dominator moves about the subordinate avoids him; when the first sleeps the second gets bolder, but still avoids the lair of his tyrant. All the same, in most cases the subordinate animal takes possession of a lair for himself, to which he retires when attacked, and the despot leaves him in peace there; but should he continue to molest him the subordinate chases him away with cries and the pretence of an attack near the entrance, whereupon the other will not persist, unless the subordinate animal has mistaken another box-lair for his own (even if the size and construction of both seem to be exactly the same).

Within a family, aggression does not make its appearance in the young males till about the end of a twelvemonth. On the other hand all the adults in a family will attack a stranger, which often creates so much confusion that the intruder escapes; sometimes the pursuers then start fighting among themselves. Even the young who have hardly yet left their parents will rush forward to meet a stranger adult and bring about its withdrawal, although they will fly from this stranger if they meet it outside their own territory. Females in milk are particularly aggressive, even with their mates. The attack usually starts with the first contact, and immediately the attacked one flees. The sight of a fleeing mouse invariably induces pursuit. The female, on the other hand, never flees; she flattens herself on the ground and cries as soon as she feels a bite: the attack then stops immediately.

This is the place to take notice of the curious submission behaviour which is found not only in mice but in all animals subject to pursuit by a despot. An attacked mouse turns and rises on its hind legs to expose its vulnerable belly; under the same circumstances the wolf offers its throat; and the attack stops at once. Tinbergen and Moynihan have criticized this idea of the submission ritual expounded by Lorenz; according to them, the exposure of vulnerable spots is merely an accidental consequence of flight: the dominated animal then turns away from its rival, which had the effect of hiding the coloured spots or bodily characteristics which when presented are the prelude to menace and attack. But to anyone who has watched mice it is impossible to agree with Tinbergen and Moynihan: it is a clear case, as Morris pointed out to them, of a well characterized posture and not an accidental one. The ritual of submission is indeed the exact opposite of the threatening and aggressive gestures, and in the case of certain fish such opposition is particularly clear: in an attitude of menace the mouth is pointed downwards towards the bottom and in submission the mouth points upwards and the tail trails on the bottom.

Men are the same here, for when they wish to surrender they also start by throwing down their arms and offering themselves defenceless to the conqueror. With the Greeks the suppliant, ἱχετηζ, was protected by Zeus, but in the Homeric age his request had to be accompanied by a series of ritual gestures that put the conquered under a moral obligation to listen to him: he grasped the victor's knees with one hand and touched his forehead with the other. This did not always prevent the thrust of the bronze sword "greedy for human flesh" into his stomach. For men are much fiercer than mice.

DOMINANCE AMONG CRICKETS

It has been maintained for a long time, I don't quite know why, that dominance and the social hierarchy are found only among the higher animals. But to those, like myself, who have raised household crickets for years, a certain amount of competition around the watering place reminds one closely of the phenomenon of dominance. And now, in 1961, Alexander has published an important paper on dominance in the field cricket. Such insects form chains of dominance in which every kind of fighting can be found. The most usual is to entangle the antennae or to hit at the anterior part of the body. If honour is is not satisfied the males jump, give the attacking cry and finally bite the conquered insect and throw it to one side; mutilation is unusual. The fight grows more intense the longer it lasts, and is also more intense between insects of near rank. Moreover a male can rise in rank through copulation, isolation, occupation of a crack, or a successful battle. The highest rank is usually obtained twelve days after the final moult, and is kept until the approach of death. Generally, but not always, it is the bigger males that dominate the smaller ones; rank is not changed by covering the eyes with varnish, by taking off all but the basal joints of the antennae, nor by fixing a ridge of cardboard on the anterior part of the thorax. But the removal of the basal joints of the antennae lowers rank and induces strange behaviour; a number of very important sense organs are found there. The despot usually gives the attacking cry when he fights, and this is often immediately followed by the retreat of the vanquished; this last insect rarely stridulates, and if he does, it is because he is about to rise in rank. Territories are defended jealously. As soon as a cricket has found a crack or has dug a lair, he moves about much less and commonly only travels in his immediate neighbourhood; in the course of these patrols he rises on his feet with his palps extended and becomes very aggressive, even with males who would rank above him on neutral ground; his song too is stronger and longer on his own territory. Finally the dominant male monopolizes the female of his choice and forbids her to the dominated insect, unless the latter insect has been able to mate with her before him, in which case the weaker one keeps her and the other leaves her alone.

Is it possible to imagine two more different animals than the cricket and the mouse? And yet dominance behaviour seems to be the same, even down to the smallest details. Is it, then, a tendency in organized matter so fundamental and so ancient that it takes no notice of the extreme divergence of phyla?

CHAPTER FIVE

MEANS OF COMMUNICATION: "LANGUAGES"

THERE IS no herd of animals so undeveloped that it does not have some form of communication and expression. Even chickens, who do not exactly shine with intelligence, have, as everyone can see for himself, a series of calls for the young—to summon them, to tell them of danger or the presence of food—and they have menacing cries and triumphant cries, and so on . . . in all some twenty calls. This is, of course, not at all comparable to what we call language in man. As Vandel so fittingly remarked, it is a matter of certain kinds of expression that in human beings would be cries. Without any previous learning, men of all times and all countries know how to utter a cry of pain which is understood as such, with just as little learning, everywhere and at all times by any being, always provided that he be human. . . .

Is that all the communication there is among animals? Not altogether. Leaving aside the famous language of the bees, which we discussed at length, birds, Amphibia and insects all use a wide range of sound combinations.

The technique of sound-recording has made so much progress over the last ten years that our knowledge of acoustic behaviour in animals is galloping forward. Research on fish, living in a much more favourable medium for sound transmission than air, has also started, but is not yet far advanced. As to mammals, we all know that they are far from dumb, but it cannot be said that sound is very important in their lives. The primates, our nearest cousins, are reasonably talkative, but do not have a true language. . . . Man alone has this distinction among the primates.

THE SONG OF INSECTS

The mechanism that produces insect sounds is very different from that of the vertebrates. In the case of the grasshoppers and

Orthoptera in general it consists of two saw-edged portions of the carapace that are rubbed together. In some cases it is the inner side of the femur and the edge of the elytra, in others the two edges of the elytra or the wings. The range of vibrations thus produced is very large and moreover some insects can make ultrasonic vibrations. Together with these sound-producers are auditory mechanisms, sometimes highly developed, with a complicated structure and often strangely placed: on the side of the thorax in crickets, on the forward tibias in the long-horned grasshoppers.

It has been known for a long time that males of the jumping Orthoptera sing to attract the females, and an old experiment of Regen's (1910) was to get a male cricket, so to speak, to call his female to the telephone! All that is necessary is to get the male to stridulate in front of a microphone and transmit the sound to a loud-speaker in another room: the female approaches it and even tries to get inside it.

Apart from the Orthoptera (grasshoppers and crickets) the cicadas are the best-known songsters in the world of insects. The Greeks have made famous their song, which bathes the mountains of Hellas: surely, they claim, the muses themselves must have taught them so divine an art: for my part I am more prosaically of the same opinion as the great Fabre, who saw nothing godlike in it, found the noise deafening and compared it, from the point of view of harmony, to the noise made by rubbing the teeth of a comb over a fingernail. I prefer, on the contrary, the cricket-on-the-hearth, who sings there as the fire dies down after the big autumn kindlings. Or again the little wood-cricket, *Nemobius*, which jumps in millions among the dead leaves, and whose subtle song, scarcely audible, wonderfully reminds one of the "frisson d'eau sur de la mousse" of the poet.

In general insect songs can be put into five categories: (1) the male summons call; (2) the female summons call; (3) the "courting song" of the male; (4) the male challenge call; (5) the sound uttered by both sexes when they are in trouble. The first two are effective over long distances. The song depends on atmospheric conditions, but in any case is only produced at certain times of the day, which vary according to the species. An American cicada suddenly starts to sing at dusk when the light has almost gone; at this moment, says Alexander, the song of millions of insects, all starting at exactly the same moment, is like a great wind blowing through the forest. As the light intensity drops, all is suddenly quiet again, after less than an hour of song.

As one hardly ever hears a male singing all alone but always in chorus, and a large chorus at that, some writers have supposed that the song attracts not only females but also other males. There is no solid proof, at least in the case of crickets; and other writers maintain the contrary: that the song of one male intimidates other males and stops them approaching. Nevertheless male cicadas of some species (*Magicicada*) can be so strongly attracted by their own song that they finish up side by side and climb one on top of the other.

But a further examination of the effect of the song on other males has shown that, according to circumstances, it may increase another male's activity or, on the other hand, reduce it, make it move towards the singer or move away from him, start him singing or stop him if he is already singing, or make him change his note. These different effects are no doubt due to some special characteristics of the song, which up to the present we have only been able to distinguish imperfectly; consequently we cannot yet clearly explain the diametrically opposite results obtained by different experimenters.

Moreover, when two males unexpectedly find themselves together their song changes in nature and turns to the easily recognized notes of the threat, which is closely related to dominance. In the case of crickets it is the dominant insect that stridulates first, most loudly and most often when a stranger cricket approaches. The dominated cricket is content to reply timidly or he may not answer at all, and before long will move off. Then the threat call gradually changes to the "come-here" call.

Female song is softer and does not seem to carry so far. But in some cases, such as that of the common grasshopper, *Chortophaga viridifasciata*, the female who has heard a first call moves towards the male and gives a brief song of reply, which leads the male to emit a second call and the female once more to reply; and so on. Alexander remarked that these alternate sound signals resemble the alternating light signals exchanged by the males and females of certain *Luciola* fireflies.

When the female gets into the neighbourhood of the male he changes his note: it becomes the "courting song" which has been so much discussed; this song does not have the faculty of orienting, or indicating the whereabouts, as does the summoning call, but, at any rate in the case of crickets, inclines the female to the act of copulation and encourages her to mount on the back of the male who then presents her with the tip of his abdomen (with crickets the female mounts the male).

Choruses. Many insects synchronize, alter or combine their song so as to make up choruses. In its simplest form this is just a matter of all the individuals breaking into song as soon as one of them starts. Slow-motion playing of recordings clearly show that if two insects are singing one is the leader and the other only starts a certain time after him. The parts cannot be exchanged and, if you stop the leader, his duet partner stops too; but stopping the second one only rarely stops the leader. Sometimes a sort of alternation can be found, when the song is in two parts. In the cases of the cricket *Orchelimum vulgare* and the cicada *Magicicada cassinii*, the first part of the song is very short, a mere "tik" and the second is a murmuring "buzz": then the insects go "tik" and "buzz" all together. It is hard to see the reason for such strange behaviour. Alexander and Moore noted that the chorus drew the males and females together much more efficiently than when the song was dispersed.

Interspecific recognition. Now there are more than 10,000 different species of singing insects: how do they recognize each other? Recordings with modern apparatus help us to understand. First of all, many insects are capable of not one, but several, stridulatory rhythms. The basic element is the utterance of one or more groups of vibrations of a definite frequency, comparable to the phonemes of human speech; a phoneme is separated from its successor by a definite interval, varying according to species, and the cadence of successive phonemes emitted is the first element in recognition. But other variations can be produced and the cricket's song is a good example: his courting song differs from his call song in that each train of impulses is less intense and more "basic", that is to say he starts and ends less abruptly; each trill of song contains twice as many impulses; finally, between two trills the male gives a short and sharp sound quite separate from the others.

All combinations are possible, according to the frequency of each trill of impulses, the cadence of the trills, the length of the phrase in relation to the interval of silence which separates it from the next phrase, or even the regularity or irregularity of this interval. Thus we already have a number of different elements whose combinations could be infinite. Nevertheless we can find within each song a "grouping of groups" which further adds to the complexity. Emissions of a given frequency, in groups separated by a certain interval, may form part of a "phrase" of a given length and be separated from the following phrase by a much longer silence. This is the case with many long-horned grassland grasshoppers, for example the *Conocephalus* family. One of them *Amblycorypha uhleri* probably produces the most complicated song found among insects. It is made up of different kinds of emission, some of which are made merely by the rubbing of a single tooth of the stridulating apparatus against the corresponding tooth of the other elytra. Within the three groups of sounds that can be noted one can also find regular rises and falls in intensity. These trills are produced in sequences that last up to a minute and a half and rarely less than 40 seconds; the differences in length are never due to an omission of part of the song but more to differences in the length of each portion. We do not know the purpose of these complicated "methods of expression". It appears that the night song has certain characteristics that distinguish it from the day song and the kind of singing is also altered when two males find themselves close together. The threat to a rival is also found in other singing insects, but the nuance is then arrived at quite simply and with much less effort. It is things such as this which make it difficult to explain the song of *Amblycorypha* from the evolutionary point of view. Another complication is found in the inheritance of sound emission: Alexander has reported that hybrids of *Nemobius Pennsylvanicus* and *N. fultoni* (another species) emitted sound with the frequency of one parent and the length of phrase of the other; the *cadence* with which the phrases were produced was intermediate between the two parents. In the case of *Amblycorypha*, which we discussed above, there is a species, *A. rotundifolia*, which has two forms, morphologically indistinguishable, but which can be separated by their different songs: moreover the two types live in different habitats with a transition zone between them where the two "species" mix. But put them in a tank and you will quickly notice that they will both set up a chorus, in which the song of the one group does not influence the song of the other. It is thus quite possible that two isolated groups have been produced within an apparently homogeneous species on the basis of differences in sound production alone.

TOADS AND FROGS

These too are vocal animals; and as they are used more and more in laboratories, being easily bred, there is no difficulty in carrying out research on their vocal apparatus. It is quite different from that of insects and closely resembles our own, except that the vocal chords of the throat are reinforced by big air sacs, sometimes of an absolutely monstrous size, which fill when the animal gives tongue; they then serve as sounding boxes. It is strange, on the other hand, to note that the *problem of hearing* in batrachians has given rise to considerable controversy, and certain people have even maintained that they are deaf and cannot hear the noise they emit. In fact they can hear, as modern research in neuro-physiology has shown; but their sense of hearing is not quite comparable to that of fish, which live entirely in the water, nor to that of terrestrial mammals, which live surrounded with air. For example, possibly some of the vibrations are not directly transmitted by the external ear, but indirectly through the body, by means of the lymphatic sacs, to the internal ear. Moreover some frogs react better when the ear drums are half immersed in water, and better results are obtained with interrupted sounds than with a continuous sound. There are very big differences according to the species concerned and one must be wary of making generalizations. Finally, outside the breeding season, which is sometimes very short, the males may behave as if the sounds no longer "interested" them.

In any case the vocal signals produced by batrachians are of a more complicated nature than those emitted by insects: not only can the frequency vary, as with insects, but also the timbre (that is the proportions of harmonies in it), the length of the individual phrases and the rate at which they are repeated. Moreover it is the repetition rate rather than the sonic frequency or the timbre that produces the most marked reactions.

As we shall see below, the call emitted by the males has many functions which have been summarized by Bogert in a recent paper.

1. Once males and females have assembled on the breeding ground (there is a strange story in connexion with this which I shall tell later on), it helps them to find a member of their own species, because in any one pond several species of batrachians can usually be found. Moreover certain females, as in the case of insects, can reply to the males' call with a special song of their own.

2. It has also been suggested that the call helps the frogs to locate the breeding pool, but, as I have already mentioned, the matter is much more complicated than that.

3. It is not impossible that the male's cry serves also to mark and maintain his individual territory.

The male's *sex-call* is the one most often heard; generally in a "chorus", as in the case of insects. It varies greatly in tone and length according to species and it is also uttered in different places, according to the creature's way of life. Certain *Hyla* (tree-frogs) croak in bushes or trees a metre or more above ground level; but there are some toads that croak away at ground level or even at the bottom of their lairs; some are half immersed with only the head above water; there are even some species which are completely aquatic and give tongue only under water. The *release cry* is given by a male when another male tries to copulate with him, mistaking him for a female, for batrachian males seize any moving object that passes near them when they are in a mating mood: male toads have been known to squeeze tench by the head to the point of suffocating them. There is, then, no real recognition of the sexes; but if the male by chance finds a female, she allows mating to take place, whereas if the creature is another male he frees himself by giving the characteristic call. In fact the vibration of the sides which accompanies the sound emission seems to be more effective in securing release than the noise itself. But tench are dumb, and that is the cause of their difficulties with toads.

Toad's territory. The male call is addressed especially to females; but in certain species at least it has the opposite effect on males. As with birds it proclaims that the area is private property, and the proprietor has no wish to receive visits from intruders. In this respect there is nothing stranger than the behaviour of the Texas cliff frog (*Syrrhophus marnocki*), studied by Jameson. These are never found at distances of less than 2 to 3 metres apart and their density is scarcely more than 8 to 9 per acre. Jameson stated that any empty territory is promptly invaded by males from the periphery. Over a period of 30 days he removed 87 males from an area of 8 acres, after marking the owners of the territory as far as possible: in the centre of the cleared area he subsequently found 46 per cent usurpers who had come in from outside, after a trip of more than 100 metres, which is much more than the distances they usually travel. Trying the opposite tactic, he released 25 males in the middle of an already occupied zone: they could not stay there and he found them on the periphery, 150 metres away. We must add that the call of *Syrrhophus* can easily be heard at more than 100 metres distance. Yet the migrants who occupied the vacant space had a territory already: why then did they leave it to take up a home deserted by a neighbour? A batrachian psychological mystery. . . .

The distress call. When caught by a predator batrachians often give a *distress call* which has a particular note and is made with the mouth open, which is exceptional, as croaking is usually done with the mouth shut. It appears that this cry is not without effect on other members of the species; in any case the respiratory rhythm changes when they hear it. Also, when you approach them without any particular precautions against being observed, frogs jump into the water, often with a special croaking sound; moreover the reverberations of the diving alone are enough to alert them all, so much so that one cannot take a step without the whole population taking shelter under the water. More often, no doubt, the first warning of approach by an observer is given by the vibration of the earth under his feet as he walks, and the chorus suddenly stops. It is the same with insects. How many times have I tried to catch Ephippigerinae, those big grasshoppers that haunt the dry meadows in August and stridulate madly on the brooms. But approach as quietly as you may, as soon as you are a metre or more away from them all noise stops. You must then remain absolutely still for several minutes for the song to start again; you generally find the singer right under your nose (for it is the same colour as the vegetation). Thus the frogs and the grasshoppers both behave in the same way, but in the case of frogs you can wave your arm without frightening them, so long as your feet do not move on the ground, whilst with the Ephippigerinae, who have better sight, you must avoid all movement.

Choirs are also found among batrachians, and deplorably noisy they are, as everyone who has lived near a pond knows. According to an old German belief every lake or pond has its old conductor frog, full of experience, who gives the note to the choir. Some naturalists have mantained the same thing, but modern research leads one to abandon this idea. The frog or toad with the most powerful voice (and the differences between individuals is very marked) is not necessarily the leader, nor does he have the socially predominant role; moreover the hierarchy is not very fixed in the case of frogs and toads. The first one to mate with a female is not necessarily the despot, but the first to get hold of her. On the other hand, what one does find is trios and duos, as Gouin noted in the case of the tree-frogs (*Hyla crucifer*) and Blair in a number of batrachians. With *Hyla crucifer*, "the first call is made by an individual emitting a single note a certain number of times. After a short silence, if he gets no reply, he gives a trill, which seems to act as a stimulus, because another individual generally replies, emitting another note. Then the two go on indefinitely with their alternate notes. If a third creature does not answer, the first one stops and gives tongue to another trill. A third may now reply on yet another note. The three will then continue each their own note and always in the same order" (Gouin).

A batrachian mystery: attachment to the breeding area. Batrachians always come back to the same pool to spawn; males and females make long journeys for this purpose. It is only a particular pond that will do, and if they are thrown into what seems to be a very suitable pond, they will climb out of it and return to their favourite place. It has been put forward on several occasions that the first males who reach the chosen zone draw the others to it by their song, but this is by no means the case, for there are even species that travel to the breeding zones in silence.

What, then, have we left that will explain this strange behaviour? Some special appearance of the place? But Anderson reports the case of a large number of the little tree-frog *Microphyla* that used to assemble in a pond some twenty metres in diameter. One spring the surrounding fields were levelled and the pond filled in, after all the area had been cleared. After a heavy rain in June the author saw some thirty frogs croaking away between the plough furrows on the very site of the former pond, and there was nothing whatever there by which they could recognize the place. To this must be added the fact that toads unfailingly make the journey to their site, even when their pond has been dried out. This is enough to eliminate the theory of "hydrotropism" (attraction to water). Savage has pointed out that frogs and toads generally make their journeys after heavy rain has soaked the whole neighbourhood and that it is quite impossible to talk of hydrotropism under such circumstances. The same author criticizes another theory according to which the frogs and toads are content to take a route along the steepest downward slope: they thus naturally get to the valleys, where the ponds are. But Boulenger let some toads out between two ponds a certain distance away and found that they chose one only, their usual one in fact. To get to it they had to traverse and climb all sorts of obstacles, particularly some hillocks and steep banks: thus the "theory of maximum slope" disappears in its turn.

Savage thought that it might be a special odour that drew the creatures, for each pond is surrounded by all sorts of plants and different species of algae grow in the water and always in different proportions. Thus it is not impossible that each pond has a special smell and that the batrachians know how to recognize it from a distance. This seems to have been proved in the case of the frog *Rana temporaria*, particularly studied by Savage. But one is a little shy of accepting this theory when one thinks of the tree-frogs that found their breeding zone even when it had been filled in some months before. . . .

THE SONG OF BIRDS

Heaven knows how many biologists have been drawn to the fascinating study of bird song! Most of all they have been attracted by the passerines (nightingales, tits and warblers, etc.) because of the variety and the beauty of their song. It quickly became apparent that, unlike everything we have seen so far in this book, bird song is not entirely innate, but is in part learnt from parents and neighbours. We can distinguish these differences thanks to some good research work done by German experimenters: their birds were raised in isolation in sound-proof rooms from the egg onwards; thus they had no opportunity of hearing any members of their own species. The Germans called them the "Caspar Hauser birds" (*Kaspar Hauser Vogel*) in memory of the famous nineteenth-century child who was raised in complete isolation, away from all human contact. It was observed that the notes emitted depended on the degree of the bird's development and, as I have said, on the presence of conspecifics. The innate development of song seems to be much the same in all the species studied; it is regulated, at least in part, by the endocrine glands. The call notes of the young gradually become, at different times according to species, the second-stage song where new themes are developed—the *subsong*, as the English call it—then a third stage arrives and finally the definitive song is reached: this does not appear until the first springtime of their life and it is characterized by the loss of several juvenile themes and by the acquisition of many new ones.

The most interesting problem is obviously the influence of other birds on the development of song. First of all we must note the appearance of a *sensitive period*, which is fairly short and lasts from the first flight until the first spring, after which the bird does not seem capable of learning to sing. Then from whom exactly does the bird take lessons? It is rather difficult to decide, in view of the mass of more or less contradictory experiments. It seems that the strongest influence is that of the bird that feeds the young, *even if it is not of the same species*. Nikolai has reported that in the case of some isolated bullfinches some phrases of song did appear, but they remained rather primitive and undeveloped until they were allowed to hear other birds, who brought them food. In another group a canary was feeding some bullfinches; a little later the bullfinches started to sing like canaries, even though they could hear males of their own species shouting their heads off in the neighbourhood. But the phenomenon is not at all simple; firstly because the themes learned may not appear in the song until some months have passed, and secondly because all Passerines are great imitators: during their sensitive period, which, after all, lasts some weeks, they may introduce into their song a whole range of tunes borrowed from other species. As other species are not found everywhere in the same proportions this means that particular localities have their own vocabularies, according to the numbers and origins of the birds. Not to mention the "family traditions"! I believe it is legitimate to use this expression about nightingales, which often nest in woods and are separated from one another for long periods. It is quite easy in recordings of song to find fairly big differences between one wood and another, and also in the themes which the singers transmit from father to son. The same thing can be seen with finches in the wild parts of Scotland; each glen has birds with their own particular "dialect". Marler remarked that these differences are strengthened by the fact that the young males learn their song from the older ones, and always come back to nest in the same spot; they have the habit of always replying to a song with a song of the same type, *provided that they know it,* which strengthens the local usage and leads to the gradual disappearance of methods of expression strange to the district. But finches have still other peculiarities: in some places, such as the Durango forest in Mexico, Marler found a finch population where no two males sang alike; so much did the songs differ one from the other that it was hard to believe that the birds belonged to the same species. . . . It must be noted that there was no question of geographic isolation; all these anarchic tenors were living in the same spot. Marler owns that for the moment he is at a loss for an explanation.

An attempt to classify animal sounds. Some aspects of bird song become clearer when they are compared with the noises made by other animals First we find a series of *sounds to do with hunger or the presence of food.* Collias even raised a young American robin (*Turdus migratorius*) that made different sounds according to the intensity of its appetite. He could keep the bird quiet for a time by giving it a worm; after a little while it began to cheep feebly; a few minutes later it made rather louder calls of one syllable, and finally it uttered a series of noisier notes of two syllables, louder than ever. Moreover, just before seizing the worm it coveted, the robin uttered a series of short notes, very piercing and rapid. These last are characteristic of the young when they see their parents: if they are out of sight there is no point in making a noise; and as soon as Collias hid himself his bird also kept quiet. On the other hand, parents give a special call when they are bringing food; when Collias gave the low whistle which in robin language means "here is some food", the young opened their beaks even if their father-feeder had in fact nothing to give them. The "food call" is well known to duck shooters, who imitate the sound the world over, and the birds are easily deceived by it. Frings has noted a curious fact in the case of gulls: if a bird finds a small piece of food it swallows it with no further ceremony, but if there is a good supply, it gives the food call, which draws the other gulls to it.

This can no doubt be compared to the "hunting call" of wolves; there are three kinds: a howling, soft and prolonged, the simple rally call; another, higher and on two notes: this means the scent is hot and must be followed; and finally a baying followed by a howl, which means the game is in sight, and is the call to action. Obviously things do not always go well and sometimes the wolves cannot afford competition from others for a restricted supply of food. Then *threatening cries* are made, as in the case of birds and mammals (a dog defending his bone), and these vocal duels are often enough to cause the adversary's withdrawal.

We cannot end this chapter on exchanges of information *within* a species without recording an extraordinary exchange *between* species: the case of the black-throated honeyguide (*Indicator indicator*). This well named bird aims to attract a human being's attention or, failing a human being, then a bear's. It likes honey, but dares not attack bees; but it knows very well how to find the wild nests. Furnished with this precious information the bird sets out to find a man and flutters around him with such strange cries and odd behaviour that it is impossible not to understand it: the bird wants to be followed. As soon as the bird sees that it has been understood, it flies towards the bees' nest, uttering cries all the way and flying back from time to time to see if it is being followed. When the man gets near the nest the bird's excitement is at its peak: it flies towards the bees then towards the man, singing at the top of its voice. It is impossible to refrain from rewarding such efforts, by uncovering and robbing the nest, and enough honey is always left in the comb to compensate the bird for its hard work. Moreover it has just been discovered that an enzyme system in the bird's intestine allows it to hydrolize and digest waxes. I wonder if it ever happens that the bird conducts its human helper not to a hollow tree but to a hive; I have not found such a thing described in literature. What happens when the bird cannot find a man? It approaches a bear, which seems to understand very well for the animal turns and follows the bird with great assiduity. The bear's cupboard love for bees is so great that Yugoslav beekeepers have to surround their hives with live electric wires in order to dissuade the plantigrades from taking the crop before they do.

Naturally one cannot resist speculating about the evolutionary mechanism that can lead to such a fantastic result. Some will lay it once more at the door of natural selection. Apart from the fact that it is too facile an explanation and one that belongs to the infantile period of biology, it is impossible to imagine that it could be proved: we are left for ever with a question. I prefer to believe that science is short, since it was born but yesterday, and in nature there are many things that we have not yet been able to explain.

The approach of an enemy. Everyone knows about the alarm cries of birds: the Frings have recorded them with the particular object of driving rooks off fields or keeping gulls away from fishing grounds; the effect is perfect and one is simultaneously freed of birds and deafened by loud speakers wired to the tape recorder. The scouts may indicate the kind of enemy to be expected by giving special cries according to whether it is approaching from the air or from the ground. The effect is to produce immobility and silence among the birds, especially among the young, which stop cheeping. It can be objected that the alarm cry itself must enable the predator to locate its prey, but several devices seem to overcome this. When the falcon approaches the habitat of the common bush tit (*Psaltriparus minimus*) all the small birds make the devil's own row for a few minutes: Grimsell remarked that the effect was to make it impossible to identify the direction from which the noise came. As to the little ground squirrels of California (*Citellus beecheyi*) it is possible to tell from their alarm cries whether a falcon, a snake or a predacious mammal has been sighted. The degree of danger is evidently shown in the cry, and above all the comparative proximity of the enemy. As one gets progressively nearer to the nest of the bird *Dumetella carolinensis* the alarm notes are short and repetitive, but when one is very near the bird starts to miaow like a cat: hence its English name, *catbird*.

It should also be noted that alarm signals of one species can be understood by members of another species, for instance seals dive into the water when cormorants give notice of danger.

Sexual life and sound signals. We may be sure that these are very numerous among birds. Some calls serve to announce the occupation of a territory, and if a rival has the audacity to reply the legitimate owner dashes to his frontier eager to chastise the intruder. All shooting men know these calls and use them to attract birds. But the same call, or one very like it, attracts the female: after she has joined the male he almost gives up singing. When several birds live near each other, though each on a different territory, a special song lets everyone know that females are approaching. Moreover in many species the female replies, and, as we shall see, it is impossible not to conclude that in monogamous species the two partners recognize each other by means of subtle individual characteristics in their exchange of song. We must also note the special call put out by the female southern house wren, *Troglodytes aedon*, which is in effect an invitation to immediate mating.

Vocal exchanges between parents and offspring are highly developed, apart from those we have already discussed in connexion with the approach of enemies. It also happens that one parent, who has gone off foraging whilst the other sits on the eggs, may announce his return whilst still out of sight, by a special call. Certain ducks (*Aix sponsa*) which nest in the fork of trees above the ground encourage their young to leave the nest by a characteristic cry. As to the young, they do not only yell their hearts out when they are hungry or frightened: they also sometimes make sounds that seem to show contentment; this is true particularly of mammals, but it also occurs among birds, such as chickens and ducks. Everyone knows the call of the hen, a cackling which will draw chickens even if it issues from a loud-speaker: the actual sight of the mother, then, is not very important; in the same way it is the chick's cries that attract the mother's attention and not the sight of her offspring. If the chick is under a sound-proof glass bell-jar, the hen will pass by, quite indifferent to the little creature, although she can see it very well.

Finally, the exchange of sound signals between members of the same species can have, in my view, a yet more curious side when its object is to synchronize the movements of a whole flock of birds.

This is so with all nocturnally migrating birds; these utter a regular series of notes, the effect of which is to rally the

Fig. 44. Brückner's experiments on the reactions of a hen and chicks.
Above. The hen can see the chick put under a glass cloche, but cannot hear its cries so takes no notice of it.
Below. The chick is hidden from the hen which can hear its cries and anxiously tries to find it. (After Tinbergen)

stragglers; in any case it is rare for a flock to take wing without giving tongue to a whole series of unvarying calls. *Do the different kinds of calls have anything in common?* Collias, after studying certain spectrograms, thought that they did. With birds and mammals all the alarm cries are sharp and sudden, whether continuous or repeated. Threatening sounds are less sharp though still sudden, like the growling of a dog; it is less well-known that sparrows under similar circumstances utter a threat cry resembling a dog's bark. *Assembly calls* made by parents are generally soft, with a low frequency, and repetitive.

All this led Darwin, not without reasòn, to draw parallels between the vocal expression of emotion in man and in animals; but, as Collias says, the vocal apparatus of birds, the syrinx, is not in the least like the vocal chords of mammals. In fact it is so different that the incontestable similarity of sounds must be attributed to a strange convergence in the phenomena of evolution. Heckett has noted something to which insufficient attention has been paid: that man has a unique vocal apparatus, even compared to the other higher primates: for instance, in non-human primates the larynx is found very near or even touching the soft palate: the human larynx is set deeper in the throat and the base of the tongue can move in the cavity thus formed. Moreover the soft palate can move freely backwards and upwards whilst in the other primates it cannot move so easily. All this is certainly connected with the very subtle modulations of the human voice. On the other hand not all emissions of sound can be called "vocal". The importance of the mouth and ears in human life is shown by the large size of the part of the brain to which they are connected. Thus Penfield and Rasmussen made some "homunculi" (models) that summarized extensive research on the critical zones by means of punctures made with very thin electrodes. Washburn (1959) reproduced the Penfield and Rasmussen models side by side with an analogous model corresponding to a monkey. The difference in the distribution of the zones is enormous.

LINGUISTIC AND SEMANTIC FACTORS

When we come to the characteristics of the message itself we must point out that with animals there is no possibility of written records. It all fades away very soon. But the ephemeral nature of the message is made up for by *frequent repetition,* which is also found in the case of man. Next, the messages are *interchangeable,* that is to say the broadcaster can as easily become a receiver. They are subjected to total feed-back; the creature hears all it puts out and this, no doubt, conditions the continuity and regularity of its calls, which are specialized; that is to say they have no common physical or biological relationship with the effect they are going to produce on the receiver. To tell the truth, it is sometimes hard to define the difference between a specialized and a non-specialized signal. The phrase "dinner's ready" has no relationship, other than a conventional one, with food; it is a specialized signal. The swollen belly of the stickleback, which will provoke the pre-copulatory pursuit of the male, is a very effective signal but one directly related to what is to follow, the deposit of eggs in the nest, therefore it is a non-specialized signal. The *semantic nature* of the signal consists in the association of each of its elements with an element in the situation with which the signal is concerned. When a gibbon finds food it gives a special call, but that does not mean that it identifies food with this cry any more than we identify the sugar we put in our coffee with the word "sugar", but there is nevertheless a close association between the two: the gibbon's call certainly has a semantic value. The signal, however, is arbitrary: that is to say nothing in the sound itself recalls the nature of the thing indicated: we call a dog *chien*, *hund*, etc., according to the countries from which we come. The relationship between these words and a particular mammal is close, but purely arbitrary. Languages are made up of *separate elements*; that is to say the elements are discontinuous and infinitely variable; the sound signals of animals can be continuous, or at least vary in a continuous and consistent way. The signal may be displaced in time, that is to say a *delay* may occur between the observation of a situation and the emission of the usual signal. A communications system can also be *open*, that is, capable of transmitting entirely new information by means of small variations in the signal. The degree of "openness" expresses, if you like, the richness of possible combinations. Finally, *duality* is one of the most abstract characteristics of signals: in a language one finds *phonemes* (the smallest elements consisting of a distinct sound emission, in practice, syllables) which have no meaning on their own, and *morphemes* (the smallest elements endowed with a meaning, that is, words). But these are found only in man and not in animals, although there is some doubt in the case of song-birds.

All these characteristics have been summarized by Heckett in a table, which I have simplified, and give opposite.

	Crickets and Grasshoppers	Dance of Bees	Display of Sticklebacks	Nightingale Song	Gibbons' Calls	Human Language	Human Music
1. Vocal auditive nature	Auditive, non-vocal	no	no	yes	yes	yes	auditive non-vocal
2. Labile nature	yes, but signal repeated.	?	?	yes	yes, (often repeated)	yes	yes
3. Interchangeability	limited	limited	no	?	yes	yes	?
4. Total feed-back	yes	?	no	yes	yes	yes	yes
5. Specialization	yes	?	partly	yes?	yes	yes	yes
6. Semanticity	no?	yes	no	perhaps	yes.	yes	no
7. Arbitrary nature	?	no	—	?	yes.	yes	—
8. Discontinuity of signals	yes?	no	?	?	yes	yes	partly
9. Delay in signalling	—	always	—	?	no	often	—
10. Openness	no	yes	no!	?	no	yes	yes
11. Duality	?	no	—	perhaps yes	—	—	—
12. Transmission by tradition	no?	definitely not	no?	partly	?	yes	yes

CHAPTER SIX

THE CONVULSIVE AND CEREMONIAL WORLD OF BIRDS

In biology the problems studied, and even the mentality of the observer, largely depend on the material chosen. Lorenz and Tinbergen were able to build their great theory of releasers precisely because their chosen animals were birds. In these creatures, in contrast to mammals, all the characteristics of behaviour are enlarged and exaggerated to the point of caricature. Nothing is more typical in this respect than the amazingly elaborate sexual displays put on by birds, accompanied as they are by cries, frantic flutterings, wild dances, and "ecstatic" postures, particularly when you compare them with mammals whose motto seems to be, "No ceremony, let's get straight to the point". Let us examine some of the complicated bird rituals, since there is not much to say about the crudity of mammals.

DO BIRDS "KISS"?

It would be easy to believe so when birds are seen to lay cheek to cheek, or when the beak of one gently caresses the other's neck.

Armstrong refers to kissing in grebes, herons, guillemots, etc. . . . The male puffin rubs his beak softly on the female's beak; they then press themselves together, breast to breast, whilst rapidly shaking their heads, after which they both make deep bows to each other. Lovesick rooks hold the loved-one's beak in their own for a while and Newman maintained that elephants in the same mood place the end of the trunk in the throat of the other.

Darwin could only compare them to savages who rub noses with a friend in the way of civility.

The male gannet, a common seabird, is a great one for bowing and can be seen with its head bent down to its feet, wings outstretched and tail up; this is generally done for the

Fig. 45. Display in seagulls. (After Tinbergen)
a and b: Beak pointed towards sky or ground.
c: Head tilted to one side during sexual display.
d: Fright attitude taken up by dominated and threatened bird on the right.

benefit of the female, but he likes the gesture so much that he will go through it a hundred times all alone on a rock. Armstrong, who greatly appreciated the funny side of ornithology, said that the sight irresistibly reminded him of two Japanese gentlemen in the middle of a busy street going through endless compliments interspersed with innumerable bows.

The wedding gift. When the tern wants to get married, he starts by catching a fish and offering it to his female, who holds it in her beak without eating it. Many ornithologists believe this to be a recognition ceremony between the spouses, and indeed the male has several females in view before he makes an offer of the fish to one of them, and some virtuous

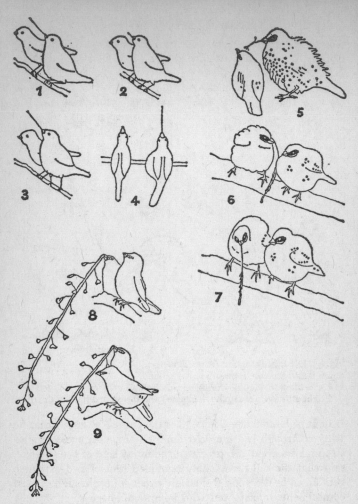

Fig. 46. The ceremony of offering a straw or twig in *Estrildinae*, Africa and the Pacific.

1 to 4 *Estrilda, Lagonosticta* and *Eudice*; note the position of the feet which are different in the three positions 1, 2 and 3.

5, 6 and 7. The same thing with *Amandava amandavs*. In 5 a male with the feathers fluffed up offers a twig to a female. In 6 and 7 females offer an object to a male.

8 and 9. The same thing with *Lagonosticta caerulescens*. Both sexes offer a piece of plant; the bird to whom the offer is made bows its head. (After Kunkel)

females have been known to return it; however this does not depress him for long, and he hurries on to offer it to another bird. At other times, the male and female each hold the fish by an end, always without eating it. It is only much later, when the female is feeding the young, that the presentation of a fish ceases to be symbolic, and that she really eats it.

Moreover females in love show a childish eagerness in demanding food that they do not eat. Several authors have pointed out a certain "childish regression" in female behaviour at the time of the pre-copulatory display; it is instructive to note that in several cases the symbolic offering is accompanied by the cry with which the male always announces his arrival

Fig. 47. A female gull (on the right) asking for food from a male, showing the special "flattened" juvenile position. (After Tinbergen)

when bringing food to the young, and here he is using it for his mate! Female gulls who had just eaten have been seen begging the male for food as soon as they came back to him, although he had stayed to guard the territory whilst they were away fishing and obviously could have nothing to give them. Moreover Noble and Wurin have shown that in the case of the blackheaded gull the demand for food, and not the actual giving of food, is an essential element in the pre-copulatory display. This shows what a poor trick it was that Mason played on a male corncrake in giving him a stuffed female. The male tried to copulate twenty-three times, and in desperation at his failure came back once with a caterpillar which he presented

to the dummy! Other birds are more practical: the male brings
a present, but the female must first yield to him; she then
shamelessly demands payment, flapping her wings like a
fledgeling.

The offer of a stone. In species where both sexes sit on the eggs,
there are many ceremonies designed to get one spouse to give
up its place to the other. An example is the great plover, which
calls softly to its mate, who then runs to replace the sitting

Fig. 48. The ceremony of changing turns between incubation partners in
Pygoscelis. (After Roberts, in Grassé, op. cit.)

bird: but just a moment! the sitting bird gets up, takes a little
stone and offers it to its mate with a bow; the other must take
it in its beak, or the bows continue but the sitting bird does not
give up its place. In other breeds (*Sala dactylatra*) stones are
offered before copulation. The male takes a small stone and
puts it in front of the female, who takes it in her beak and puts
it down a little distance away; this can go on for one or two

hours. Gannets offer bits of seaweed to each other. The male *Bombycilla cedrorum*, seated on a branch, gives his female a shiny berry, and she gives it back to him several times in succession. With herons, as we shall see, a stick is used.

The great dances. We must now consider cases where social stimulation is more marked. Up to the present we have considered only isolated couples. But with many birds, these displays are performed by groups, often very numerous, in well-defined zones that are used again and again for many years. The dance area is generally fiercely defended against any competitor and is kept clean and free from weeds. The birds' revels are so numerous and so often repeated that the soil is sometimes quite bare. In the Netherlands the sandpipers collect in the fields and mark out a number of circular areas about six metres in diameter. In the middle of the flowering fields, these look very strange to anyone unfamiliar with the habits of these birds. According to Armstrong, the sandpipers remain faithful to their dance floors and never abandon them, even if a road is built right across one of them. This attachment to the dance floor (I almost said "to the sacred grove") is very widespread. Beebe tells us how an old Dyak chief took him once to a pheasant's (*Argusianus argus*) dance area, telling him that within the memory of man they had always used that same place. Birds of paradise make their displays on the same tree every year. Gross tells the sad story of a species of grouse on the road to extinction, so much so that only one was left in a huge area; every year he went to the ancestral assembly zone, made sacred by the passage of time, a place that once had happily bristled with grouse; but until his death no bird ever came to join him there. . . .

The dances themselves are the very height of strangeness. Armstrong describes those of the sandpiper very amusingly. They have been termed fights and war dances, but they are rather simulated fights, a display. The males rush towards one another, then suddenly stop, their feet spread out, the head bowed very low and the feathers on end. This lasts a few seconds, then the bird revives and either dashes off in another direction, or, on the other hand, slowly drops to earth as if

Fig. 49. Fighting attitudes in the sandpiper (*Philomachus pugnax*), after Lindeman.

a. Dominating attitude.
b. During combat preliminaries this attitude follows a.
c. Attitude of male in front of a female.
d. A phase of the fight.
e. During the fight the beak is often pointed at the ground. This, no doubt, is a "substitution movement".

punctured, and finally subsides (Armstrong). From time to time the sandpipers jump into the air or circle round one another. At the height of the excitement the mass of males who have been rushing at each other stop and spring up with their feathers all rumpled; they present an odd and dreamlike spectacle. The females arrive in the middle of all this, appearing perfectly calm: a female approaches the male of her choice, by now fixed, rigid, "ecstatic", in the almost baroque posture of the sexual display they adopt, and touches his head or fondles his feathers with her beak; it is her request for mating. In general they are able to mate without interference from the other males.

In the case of certain American species of *Centrocercus* (a bird fairly close to the pheasant), the dance area can be 200 metres broad and more than 600 metres long, with 400 cocks moving around inside it. They seldom get closer than 10 metres from each other, always adopting the fighting posture, but here it often turns into a real fight. There are several areas of 5 to 6 square metres kept for the females: they must stay in these areas until they decide to mate. Moreover they are under the control of several "officers", among whom are a "master cock", a "sub-cock" and several "guards". These titles are related to the sexual privileges they enjoy. The chief cock spends a lot of time in display, advancing three or four steps, the head thrown back, the tail and wings widely spread and the chest inflated; he then suddenly collapses, uttering a short call. After this he mates up to twenty times in a morning; it is only when he is satiated that the sub-cock has his turn, more or less generously. As to the unfortunate guards, they must watch for an opportunity, when their superiors are inattentive, to satisfy their appetites clandestinely.

DANCES OF MEN AND BIRDS

Obviously all this brings us almost automatically to the dances of savages, and Armstrong has not failed to note their resemblances. It is quite enough to follow, step by step, the descriptions of explorers such as Nelson (1887) and Brandt (1943), who witnessed the dances of the Canadian crane. Nelson wrote,

"On the 18th of May I was much amused by the behaviour of two cranes (*Grus canadensis*) which were displaying near by. The first was not long alone and was soon joined by another emitting a deep note at short intervals. . . . The two birds then started making a series of calls in rapid succession. Suddenly the second to arrive, a male, turned towards the female and made a deep bow, his head almost touching the soil: he ended by jumping quickly into the air. Another pirouette brought him face to face with the object of his desire, to make another yet deeper bow, his wings hanging down on each side. She replied to this by another bow and they both tried to outdo each other with steps and sudden jumps, mixed with ceremonial bows, all gravely comic. The couple stayed still occasionally, bending to the right, then to the left . . . then a whole series of jumps and glissades took place, resembling the steps of the most ridiculous minuet one could imagine."

Two of the birds Brandt (1943) disturbed began to dance and the two young Eskimos who were with the explorer started to accompany them by striking an arrow against the leather of the kayak. "Then the two young men sang the 'crane song', accompanied by the beat of their improvised drums. As they sang the two big birds continued their dance, perfectly in time with the savage and strange music. When the young men beat the drum more quickly the dancers increased their speed, or slowed down, following the beat."

As I have already said, the behaviour of birds can be compared with that of man in more ways than one. Nothing is commoner, for example, than the detached and apparently indifferent attitude of the female. In the case of the Wanderenko of East Africa, the women quietly smokes a cigarette, her hands behind her back, and watches unmoved the dance of the warriors who want to charm her. In New Ireland, the young girl brutally repulses the masked dancers, who nevertheless are giving of their best; but still, at the end, she does not fail to make a sign to one of them. With the Amazon Itaogapuk, the Rio Yapura Tsaloa, and some Indo-Chinese tribes, the couples dance together before retiring two by two into the forest: they are only imitating the bird *Manacus manacus vitellinus*. The round dance, where the dancers follow each other in a big circle, so widespread among men, is not unknown with turkeys, which may perform it with a tree as centre. Avocets do a similar thing. The male of the rosy pastor (*Pastor roseus*) circles the female with hurried steps, wings and tail trembling, the feathers of the throat and crest raised, singing energetically; the female at first does not make much noise, but soon joins in the song and starts to follow the male; then both turn more and more quickly: this dance is known to and practised by the Caucasian mountaineers. The Kundun have a dance in which the man bounds round the motionless woman; on the other hand in the *lezghinka* the man and woman both circle rapidly round each other.

The audience merely plays the part of spectator, or accompanist, beating hands and feet, as with the grey goose and some species of albatross. One may see six to eight of these geese in a circle ten metres in diameter: a male in the centre is doing his utmost while the audience accompanies him, though not invariably, with a muffled and interrupted song.

We know that many primitive peoples use the dance as a veritable drug, and collapse in an almost cataleptic state after indulging in it for hours. It appears that birds push themselves to the same limits, the "point of ecstasy", as Armstrong put it. American grouse in this state will let themselves be captured by coyotes, without resistance: in the same way it used to be possible to capture bushmen, accused of some offence, when they were under the influence of their special dance rituals.

To this we must add that since time immemorial men have known, admired and imitated the dances of birds, as in the crane dance, in which the bird's gait is imitated. Theseus, on his return from Crete, performed such a dance with the young people he had rescued from the Minotaur. The Jivaros similarly imitate the cock-of-the-rock (*Rupicola*) and the Tchouktchis the sandpiper (*Philomachus pugnax*), the Australian bushmen the emu and the Tarahumars the turkey. Even European dances sometimes bear traces of this winged origin; in Bavaria one of the figures of the *Schuhplatter*, the *Nachsteigen*, imitates the sexual display of the mountain cock: one of the dancers jumps over his partner clicking his tongue and clapping his hands, then strikes the soil with one hand and bounds towards his partner with his arms wide open, or else lets them droop by his side, bending so that they touch the ground. As birds dance, so do men.

HOW DO THEY KNOW EACH OTHER?

If the stimuli that lead to mating are usually plainly visible (even in the case of man) this is by no means the case with those characteristics serving the purposes of neighbour recognition. They are quite inconspicuous and in many cases we have not yet learnt to recognize them. For example, black-headed gulls just know their near neighbours but, according to Lorenz, in some small colonies of crows every bird knows all the others. As with the tern everything depends on minute differences of behaviour; but these differences must be well-defined, for a pintail (*Anas acuta*) will recognize another three metres away, according to Hochbaum; and the robin will do the same at more than thirty metres, even if its fellow is hidden in the heart of a bush. We do not know how, except in a few cases where it all appears to depend, as with man, on the features of the head, I nearly said face. Lorenz, quoting Heinroth, tells how a swan attacked its partner, whose head was in the water, and did not discover its mistake until the creature brought its head up. Naturally the characteristics of the song play a part, as we have seen: some passerine birds (*Lepidocolaptes*) can recognize the voice of their mates as they swoop down on to a tree trunk, singing with a flock of their congeners.

HOW CAN ONE INTEREST THE FEMALES?

If you are an American heron (*Nycticorax hoactli*) this is what you must do (but of course you would know it from birth, or thereabouts): break off a twig and hold it in your beak whilst clacking the bill and alternately raising and lowering the head. After which the twig may be rejected or accepted for the nest platform. Besides the above, you might lower your beak almost to the level of your pink feet, then lift up a foot and the beak at the same time, doing it all some eight or ten times a minute. The European heron behaves in almost the same way, except that the pink feet of the male, which he displays to advantage, do not seem to be as important as one might imagine, since part of his performance takes place at night. On the other hand a female whose feet are not yet pink is driven away. Finally, if the birds' ears are blocked, the nuptial ceremony is much disturbed: sound, then, is more important than sight.

A heron also offers a twig when it wants to replace its partner on the nest. Sometimes one bird, as for example the swan, so exactly repeats the actions of its partner—neck bent back, then bent forward, raising of the wings, etc.—that, as Selous said, you might be seeing the bird's movements reflected in a mirror. But there is no doubt that the strangest thing is the welcoming ceremony of pelicans. The bird comes down near the nest, then, with its long beak pointed towards the sky, it slowly waves the beak from right to left like a flag; meanwhile the sitting bird lowers its beak on to the nest, half opens its wings and gives a guttural "chuck". All of a sudden the two birds start to clean their feathers with their beaks as if they do not know where to look. It is only after all this that the sitting bird will give way to its mate.

CONTAGIOUS FRENZY

This is a behavioural pattern very common in birds: ritual positions and frenzies will often spread like wildfire through a whole colony; it is a typically social phenomenon, but one particularly characteristic of the bird world, that is to say it has the automatic behaviour and "exaggeration" that so often mark avian actions. Certain ducks, for instance, make their display on a pond; but soon the interest they have in their neighbour's behaviour overrides the strictly sexual character-istics of the display: and while the females wait, the curves, dives and beatings of wings whip the water into a veritable froth.

Many other cases are known where birds show an astonishing synchronization of action: when a flock of starlings suddenly changes direction; when a colony of squalling gulls is seized by an abrupt "wave of utter silence" before the whole flock takes off as one bird. But, I repeat, we are in the world of birds,

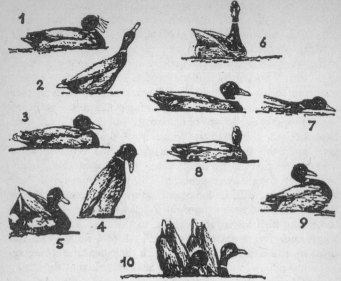

Fig. 50. Display in drakes, after Ramsey but with different drawings

1. Shaking the beak.
2. Beak pointed upwards.
3. Shaking the tail.
4. Grunt-whistle attitude.
5. Head up, tail up.
6. Drake turning towards duck.
7. Beak skimming the surface of the water.
8. Drake turning head (and showing green collar).
9. Drake puffing itself up.
10. Fore and aft balancing.

where everything is quickly pushed to its limits. If a penguin, its beak raised, stiffens "in ecstasy" and gives its rallying call, hundreds of others will follow its example; the same thing is found in terns: a captive tern, isolated in a cage, will sit on imaginary eggs if it sees other terns outside occupied in that way. A pair of Florida jays cannot continue to build their nest if they see a flock of their own breed calling and flying over them: they fly off at once, take part in the chorus and only start to work again later on.

Fig. 51. Attitudes of the bee-eater (after Koenig)
a. Bath in the rain, feathers fluffed out, beak open, eyes closed.
b. Sun bath.
c. Another position adopted during rain.

Then there is the case of the heron, noted by Lorenz, which in digesting a meal adopts the characteristic lethargic attitude of the well-fed bird, seeing which, many of its companions, though still hungry, stop eating and cannot resist imitating the example before them!

THE INFLUENCE OF SOCIAL STIMULATION ON THE DEVELOPMENT OF THE YOUNG

Observations in this realm are by no means complete, and are even contradictory on some points. It does appear, however, that the presence of congeners (or even a colony of birds of another species) notably speeds up development. This is strikingly illustrated by the fact that, in the same colony of

sea-birds, those in the centre and those on the periphery reach
maturity at different speeds. Differences between flocks can
also be noted. Behle found that there were some twenty
colonies of pelicans on an island in the Great Salt Lake and
that within each colony development of the young was homogeneous, but that there were considerable differences between
different colonies. This is a general phenomenon and has also
been seen in terns, where, moreover, big flocks do better than
small ones. It is above all Darling and his followers to whom we
owe the observations on social stimulation. Several naturalists
have criticized him, but their arguments are not always very
convincing: modifications of behaviour due to the presence of
one or more companions is so marked that physiologists have
observed them not only as group effects (see above, particularly
in the case of mice), but also in direct experiments with birds.

For instance Berry raised in a paddock a good number of
geese of different species, which were in no hurry to nest; he
then conceived the idea of giving them artificial nests and after
doing so almost immediately got eggs, and in many cases
young were reared. The following year he got the same results,
a considerable acceleration in regard to nesting compared to
the control birds without artificial nests. The same thing
happens with parrots.

Emlen and Lorenz implanted some testosterone tablets in two
Phasianidae (*Lophostyx californica*), which released copulation
behaviour the second day following treatment. Now eight other
birds raised with the first pair, but having no experimental
treatment, also showed the copulatory activity ten days or a
fortnight after the start of the experiment *and more than two
months before the normal date*.

We can no longer doubt the influence of the group on the
bird; Darling's hypothesis is thus much reinforced.

SOCIAL CO-OPERATION

There is more than a mere passive imitation of conspecifics: in
some cases the colony actively fosters the survival of the
individual. For example, according to Palmer, when food is
short among terns, many of the young die; under these circum-

stances parent birds who have lost their own young give what food they can find to the nestlings of other birds, which is never done under normal circumstances. It even seems that if conditions get still worse, the colony will select the most vigorous nestlings for feeding, allowing the weaklings to die.

Many birds help each other to make nests and offspring often help their parents with the next nest: this happens with a number of species with various tongue-twisting names that I will spare you, but also with swallows. Social co-operation is certainly found at its maximum in those truly social species that do everything in common, nesting, care of the young and defence of territory. There are some species which would repay study with enthralling information, but unfortunately such studies have hardly started and are far less advanced than those on bees and ants. Particular examples are the sociable weaver birds of South Africa whose enormous nests, several metres in circumference, at times break the branches of the supporting trees with their weight. The American cuckoos (of the *Crotophaga* genus) also have a common nest. The eggs of several females are found together in a depression in the ground and all the population sits on them in turn; feeding the young is also done co-operatively. Evidently the social influence of conspecifics, such a strong force in birds, is here exercised to its full extent. However, we are once again too ignorant of what goes on in nests held in common to be able to say anything more.

THE HIERARCHY OR PECKING ORDER

Flocks of birds (and, as we shall see later, groups of mammals and insects), if one can believe Alexander, are by no means unorganized hordes where each individual behaves as chance may decree. Quite the contrary is the case: there is a very strict hierarchical order of precedence where a despot often rules with a rod of iron. Carpenter has given an excellent definition of dominance in birds: "An individual is said to be dominant in relation to another when it has priority in feeding, sex and movement and when it exceed other individuals in aggressiveness and the power to control the group." The study of

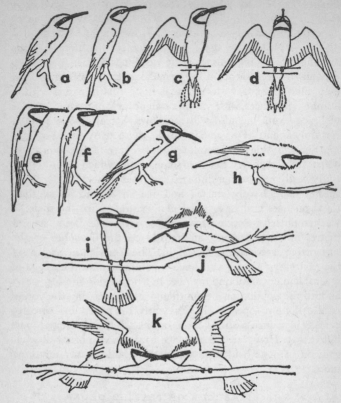

Fig. 52. More bee-eater attitudes (after Koenig).
a. Normal position.
b. About to call.
c. Welcome to a congener coming from the right.
d. Welcome to a congener coming from straight ahead.
e. Display attitude, the body rigid, see also f and g.
h. Female, with closed eyes, offers herself to the display.
i. and j. Threats.
k. Fight.

dominance is complicated by the fact that it can be manifested between different species and, here too, not by chance. For instance, in mixed flocks of passerines feeding in the winter fields the great tom tit dominates the blue tit, which in its turn dominates the marsh tit (*Parus palustris*). In penguin rookeries where several species live together the Adélie penguin (*Pygoscelis adeliae*) is driven out by other species; the blue goose dominates the Canada goose, etc. . . .

THE PECKING ORDER OF CHICKENS

The chicken, apparently such an ordinary bird, has become famous since Schjelderup-Ebbe's work. This able Norwegian scientist had the idea of counting the pecks the young cocks so liberally give each other and kept a record for each bird: this led him to the concept of the peck-order, the alpha bird pecking all the others and receiving no pecks while the omega bird is continually attacked, even to death, without defending itself; that is pecking order or, in other words, dominance.

From being simply a theory about chickens it was later found to be of very general application and became famous, as we all know. Guhl, at the University of Kansas, has just renewed the study of social order and its development in chickens. Three days after hatching no very definite reaction is seen, apart from flight from any moving object, when the chick takes refuge under its mother's wing. After a week the chicks start to run here and there with their wings spread out; during the second week pretend combats take place: two chicks jump at each other like adults about to fight, but no pecks are exchanged and the disturbances do not last long. Between the fifth and sixth weeks rather more serious encounters occur, with pecking, but the antagonists do not put their hearts into it; one of the combatants may then retire, but will return a little later and peck his opponent. True dominance with a definite peck-order is reached a little later, but it is difficult to fix the precise age: much depends on outside conditions, the kind of flock, etc. . . . At the same time, it appears, chickens recognize members of their own flock and give strangers a bad reception; Smith maintains that Leghorns are capable of making such distinctions from the age of ten days.

Pullets are much less aggressive than males; the young cocks may peck them from time to time but at full sexual maturity the cocks learn the facts of life and no longer attack the hens. The females peck each other; they belong to another hierarchy, distinct from that of the males; but whilst inter-male dominance develops towards the seventh week of life inter-female dominance does not form until around the ninth. Moreover, the peck-order is not rigid and fixed; revolts can take place in the ranks. These are due, no doubt, to individual differences in the speed of development; sometimes birds can regain lost ground. Such revolts can be assisted by temporary isolation of certain members of the flock, which allows them to recover from the stress of continually being pecked. It is quite possible to isolate pullets from birth and only bring them together when the control birds, brought up in common, have found their peck-order. But the newly grouped birds very quickly establish their own peck-order, which shows that mutual stimuli between very young chicks is by no means necessary to this end. We must also note that birds brought up in isolation are much more aggressive than those raised in a group. In certain groups there is found, in effect, a degree of social inertia (Guhl), or, to use a better term, *habituation* to social contacts, the general effect of which is to reduce the number of pecks per individual.

It is strange to note that the ingestion of male hormones by young cocks has but little effect on dominance, whilst the use of female hormone on them tends to make them more "phleg-matic", meaning that they avoid quarrels and giving peck for peck. Exactly similar results are obtained with pullets: those getting male hormone become slightly more dominant, but the difference from the untreated controls is very slight; female hormone makes them decidedly more "submissive".

SEXUAL BEHAVIOUR AND THE PECK-ORDER

A confusing complication is that there is no relationship between *the number of times a chicken is covered by the cock and her rank in the female peck-order*. Some writers assure us that high-ranking hens are covered more frequently, others say less frequently. Nevertheless, these last maintain that although the hens mate less often than their subordinates, they often lay more eggs and have more young. Allee found no relationship between the sexual activity of the cocks and their social rank in the male hierarchy except that in general the alpha animal

Fig. 53. Attitudes expressing dominance and submission in chickens, after Foreman and Allee.
a. Dominating position, head straight, feathers slightly raised.
b. A cock's dominating position (note the half-bent legs).
c. Very aggressive position.
d. This position in a hen, with legs half-flexed, shows that she has been accepted by the flock and is not particularly aggressive; the opposite of that suggested by a.
e. Attitude indicating submission.
f. Sexual attitude (invitation to mate).

mates alone and stops all the others doing so; he is particularly watchful over beta and throws himself on beta with fury if the latter shows any sign of mating. But the strange thing is that he is much more tolerant to the pariahs at the bottom of the list and may even occasionally allow omega to push him aside as he starts to mount a hen and let himself be replaced by this lowly creature. Guhl claims that cocks freed from long domination by a despot have very little sexual activity with females. He has even spoken of "psychological castration". We should note that, according to Lorenz, with jackdaws the partner lower in the scale at the time of the "betrothal" of a couple is immediately raised to the level of its future spouse by the other birds, who seem quickly to become aware of the happy event and to respect the affianced bird much more.

UNEXPECTED ABILITY IN CHICKENS

An experiment may be made in a flock of hens with an established peck-order by taking some out and putting them in another group, and then, after a few days, bringing them back to the original flock. A hen may thus have a number of different ranks as she goes through various groups. For instance, in certain experiments a hen had the following successive rankings in five different groups: 2, 6, 2, 4, 7; another in these same groups was placed in the rankings 1, 5, 1, 5 and 6. The *non-resident birds need only pass an hour a day* in the new groups for the birds to recognize them, and know their rank. How is it that birds of such mediocre intelligence as chickens can so clearly distinguish a congener without mistake among twenty-seven other chickens and know its rank as well?

Moreover, different breeds of chickens are not of the same social rank when mixed one with the other. For example, Leghorns regularly dominate all others and bantams are always at the bottom of the social ladder.

TERRITORY OF BIRDS

Birds, like mammals and some insects, claim territories, as we have seen, and jealously defend them against intrusion from strangers. They know the boundaries well and often patrol their territory in every direction: mammals mark the boundaries with various odoriferous secretions, often urine. However, at least in the case of birds, there are several kinds of territory. Noble proposed to call "territory" any area especially defended, and such a definition would do were it not for the great variety of defence reactions. An essential characteristic, in any case, is opposition to the entry of congeners, and of congeners only: for example, a stork will defend the nest area against the entry of other storks but will allow a sparrow to build in the edge of its own nest. If food is abundant, storks in Africa allow other couples around them; but they drive them out if scarcity threatens. It is usually the male that undertakes the defence of the territory, though on occasion both sexes may co-operate. But it is always the male who traces the boundaries; the female learns them, sometimes quickly, sometimes in a week as with sparrows, or possibly not at all. Sometimes the male encourages his pupil by means of a few pecks!

After the eggs hatch the defence of the territory becomes more important, even though some birds leave the strict territorial area to look for food outside while others carve out a bigger space than they need for their subsistence. It should not be thought that a territory must necessarily have a nest in it: for instance with ducks the two may be separated by more than a kilometre and the territory only be used for the sexual display; no rival dares intrude. Moreover, in general the territory seems to be needed more for sexual purposes than for hunting for food. The classic example, to leave the bird world for a moment, is that of the Eskimo dogs, reported by Tinbergen. In the prepuberty stage the puppies run over the adult male territory although they are often chased off after more or less severe punishment. But after puberty they dare not do any such thing. One of the reasons for the defence of territory is that it is an area where the couple can mate undisturbed by interference from rivals. As Armstrong roundly points out, good Englishman that he is, birds are very particular about observing the decencies.

The males of some pheasants, of grouse, canaries and pigeons, cannot bear to see another male mating, and attacks under these circumstances are so numerous (above all in farmyard birds under crowded conditions), that over-all fertility can fall considerably. Nevertheless, even in crowded colonies, a couple will mate undisturbed *on its own nest and in its own territory*, but bad luck to it if it tries to do so outside it! It must also be noted that many monogamous males guard the virtue of their flighty females very strictly and will not allow them on to the territory of a too enterprising neighbour. "Territorialism" is so strongly developed and so inflexible that Armstrong supposes it may well be the reason for the male birds' highly developed sexual display: is it not designed above all to overcome the female's repugnance to entering a strange territory?

But in bird colonies there is a much more down-to-earth reason for the defence of territory: it is that birds are terrible thieves and try to steal nesting materials from each other. All penguins think of is stealing stones from another's nest. and an unguarded nest disappears under your very eyes. According to Chapman, pelicans about to commit such an offence have a furtive air that renders them conspicuous from afar. If cormorants leave their nest unguarded for an instant, which usually they never do, gulls throw themselves on it and destroy the eggs.

A male defends his territory with great ardour, even if his rank is below that of the attacker, and for this reason males have been held to be irremovable. In fact, though, if the intruder is very aggressive he can drive out the legitimate owner. Black-headed gulls push things to such lengths that after invading a stranger's territory the intruder will try to copulate with the owner's wife in the presence of the husband. If a stranger warrior thus invades the nest of an already mated female cormorant, she will rigorously defend herself, then passively oppose his advances; this so annoys the assailant that he ends by pushing her out of the nest.

REACTIONS TO WOUNDED OR ABNORMAL BIRDS

Some birds give no help to the wounded, or even finish them off. Penguins take no notice of them. Rooks and magpies reply to their distress calls by flying to help them, uttering a special cry which draws all the flock. In the case of jackdaws help is given even to yesterday's enemy, although crows will take advantage of an enemy's misfortune to kill him. With jays and some others, if the wounded bird struggles its fellow birds get very agitated; but if it lies still the others will circle prudently round it. Terns will fly round a wounded bird: if it struggles they will scream as they circle, if it moves only feebly they will fly silently, dispersing when it becomes still; should there be much blood they will despatch their injured comrade.

In the case of gulls a sudden death will lead to the silent dispersal of the entire flock. Flocks of jackdaws, as everyone knows, get most agitated when one of their number is missing; they all set on any animal whatsoever that carries one of them off, or simply passes by carrying a black object, and will even attack another jackdaw if it is carrying a black feather in its beak. On the other hand you can take a nestling from the nest without causing any trouble, provided its feathers have not yet grown.

Abnormal birds get different treatment according to the species concerned; any mutant showing a considerable difference is attacked by penguins; in other species female albinos greatly excite the males.

However, cases are reported of help given to the wounded and infirm. For instance, a healthy blind pelican has been seen in a pelican flock: it was unable to fish but its conspecifics brought it enough food to live on. The same sort of thing has been seen in rookeries, where a blind, or even a wounded, bird has been helped. It seems that these unfit or wounded birds regress to juvenile behaviour and ask for food in the same way as the young do from their parents.

BENEVOLENCE AMONG BIRDS

There is an even stranger manifestation of inter-species association: the *helpers*, that is, birds of the same, or other, species that help others in some way or other. Skutch has recently written a general review of the subject that has set me thinking. The most usual method of help is to warn another creature by means of various cries; a bird reacts to the alarm call of its own or other species; often even mammals may benefit, as we have seen above, from the alarm calls of birds habitually associated with them. However, the giving of food by another species is fairly frequent too: crows have been seen to pass food through the bars of a cage to a black vulture. And is it true that a red cardinal bird has been seen feeding a goldfish? It might well be that the bird had lost its young and the large open throat of the fish was the stimulus needed to release the feeding mechanism. But herons are more treacherous; according to Lowell they give food to fish with the idea of luring and thus capturing them more easily.

Interspecific relationships. Birds sometimes seek strange places for their nests—driven, Mayaud supposes, by their need for security, but this has not been proved. Small passerines occasionally nest against or within the nest walls of eagles or fishing vultures (which, it should be noted, do not eat passerines); thrushes do not fear the neighbourhood of a falcon's nest, in fact they even look for one. We well know that man's houses do not alarm swallows and storks. Stranger still are the *associations between birds and aggressive insects.* Some Asiatic species even nest inside the nest of tree ants, from whom they demand not only lodging but board as well, although, in general, these ants are far from hospitable! The attraction ants have for birds is strange: for example, what is this *anting* that so many of the birds in France indulge in? It consists in wallowing in the nest of the ant *Formica rufa* with wings spread; sometimes the bird even takes ants in its beak and puts them under its wing; the ants become very aggressive, bite as much as possible and spray formic acid around: this does not prevent the birds showing every sign of complete contentment. A large number of birds live among wasp nests. The reverse can also happen and wasps establish themselves on birds' nests as soon as they are built (for example, on the nest of *Ploceus sakalava* of Malagasy). And everything seems to run quite smoothly.

HELP IN RAISING THE YOUNG

In at least twenty species of birds the young adults show a more or less persistent attachment to the nest, which may well be the beginning of social behaviour. In the case of certain troglodytes, the immature adults are generally driven away as soon as the second sitting of eggs is started. But it can happen that these "adolescents" "cling" to the nest and stay attached to it for some time; for instance, they may bring twigs to reinforce it, sometimes playing with them as if not really knowing what to do with the material. They bring food, carry excrement away and try to make themselves useful in various ways; sometimes they themselves ask the parents for food, undergoing a king of regression to the infantile state. It can also happen that a pair of birds, fully occupied in raising their own young, will look after the nestlings of a neighbour who has abandoned them, perhaps shot by someone. Skutch noted an extreme case, that of an immature *Sialia* who took on the task of feeding fifteen nestlings of half a dozen different species.

Perhaps all this strange behaviour can be explained by the abnormal force of "parental pulsion" which obliges certain birds to react by feeding any sort of nestling which opens its beak and cries. It is the most likely hypothesis, but how then can we explain Skutch's observation that *some birds neglect their own nests* to care for those of other species?

How is this interspecific help received by the parents and legitimate occupiers of the nest? Badly at times, and the intruder may be driven away, unless the latter turns out the parents in its determination to be useful at any price! At other times, things go more smoothly, but sometimes the parents take the food from the beak of the helper and do not allow him actually to give the food to the young. The peak achievement in such co-operative action is reached with those birds which lay their eggs in the nests of other species and all sit on them in turn (see page 212, social co-operation).

THE LIFE OF A BIRD COMMUNITY: A SUMMARY

It is unfortunate that we are unable to write a paper on the social cuckoos of America or the red finches of South Africa; the literature on the subject is too scarce. It would be fascinating to compare the work done co-operatively (which is the supreme test of social life) in the respective cases of bees and birds. We must be content to wait for such work to be done. Nevertheless it is worth while trying to give a short account of social life in a big rookery of emperor penguins; some young French workers have recently collected some first-class information on the subject.

THE AMAZING SUCCESS OF THE EMPEROR PENGUIN

Sapin-Jalouste and Prévost have recently given us some data on the strange life of the emperor penguins, those remarkable inhabitants of the South Pole. And when I say "remarkable" I am particularly thinking of them in comparison with higher animals trying to live in such a terrible climate. It has to be seen to be believed. Judge for yourself: the wind first of all is tremendous (the maximum is 70 *metres per second*) and calculations show that 20,000 tons of snow are blown over each square metre of the coast of Terre Adélie each year. The temperature runs from − 10°C to − 33°C; and they choose the hardest part of the winter for incubating and hatching their eggs. The blizzard carries away their calories at such a rate that Jalouste tried to calculate what their "refrigeration power factor" would be for calm weather and reached the astonishing figure of − 180°C. The emperor penguins owe their names to their considerable size: 1·44 metres high and 1·32 metres round the chest. Their weight is impressive too, 26 to 42 kilos. In spite of their vigour and the thick layer of fat that more or less insulates them, these penguins cannot live alone; no doubt that is why they are so strongly social and why they have developed a perfected social thermo-regulation system which we find elsewhere only in the case of bees. This temperature regulation is achieved by means of the "tortoise", a term that derives from the appearance of the birds tightly pressed together.

When the temperature gets disagreeably low, even for penguins, this is what they do: two or three hundred birds form a group, pressing themselves together into an absolutely regular circle which slowly turns around its centre; thus each "tortoise" leaves perfectly regular concentric tracks in the snow. When the eggs are being incubated the groups get bigger and more irregular and may contain five or six hundred birds. This "tortoise" is not immobile either, but slowly and constantly proceeds to leeward. Prévost thinks this displacement is due to the fact that the birds most exposed to the wind slide along the flanks of the mass. After a blizzard lasting 36 to 48 hours, Prévost found a movement of from 100 to 200 metres. When the storm is over the birds disperse.

Is this social thermo-regulation effective? The young French research workers were ready to try anything and set about checking this by taking the temperatures of the birds. This is a task full of risks and needs two tough young men and a good number of spare thermometers, for though penguins are peaceful birds, some are at first quite astounded by the procedure and then indignant—when their powerful wings can send workers and thermometers flying. But the results obtained make the trouble worth while: in fact at an atmospheric temperature of −19°C the birds in the massed group show a temperature at the centre of 35–36°C; isolated birds show 37°C.

At this period the birds had already fasted for two months. Under such conditions they "burn up" a considerable quantity of material when isolated; to conserve their resources they try to get into the nearest "tortoise" as quickly as possible. If they are kept isolated, they lose weight more rapidly: a hundred or so grams *per day* in the "tortoise" against 200 grams or more when isolated.

However, penguins do not spend their whole time in the "tortoises"; when there is no wind and the weather is calm and and the temperature mild, −10°C for example, display starts. The bird, hardly a fighter, moves slowly and never loses its calm. The male approaches the female after having exchanged a "courting song" with her. Then, according to Prévost, "the two birds, facing each other or not, have their heads inclined downwards, the neck forming a cross ring. If they are far away they approach each other with a characteristic balancing walk. With the neck a little dilated they stand still in front of each other, shifting their weight from foot to foot, the body bent. The head is slowly lifted to the sky, the base of the neck dilates more and more, causing a slight fluffing up of the feathers . . . The two birds may then either stay pressed together, breast to breast, or remain slightly apart. Their immobility is almost complete; their eyelids, half closed, may then blink a little; their visual and aural awareness of others around them seems to be greatly reduced. This sort of ecstasy appears to require considerable physical effort and is always of short duration: it often ends by simultaneous deglutition in the two partners, the body then becoming completely relaxed. In other cases it finishes with a sideways movement of the head, accompanied by a growl of anger."

There are variations in this curious display between courting

Fig. 54. Display in penguins.
a. Walking attitudes; on the left, a penguin calling.
b. Emperor penguin; 1, 2, 3 and 4, different attitudes during display (after Prévost).
c. The "sharpening" attitude in Adélie penguins: the two birds mutually rub their beaks as if they wanted to sharpen them (after Sapin-Jaloustre).
d. Ecstatic attitude in penguins (after Sapin-Jaloustre and Richdale).
e. Adélie penguin carrying a stone to its nest (after Sapin-Jaloustre).
g. Two stages in the distribution of food to the young (after Sapin-Jaloustre).

males and females on the one hand and mates on the other; in the last case it often precedes the exchange of the egg, for emperor penguins share the work of incubating the egg between the male and female in turn. The solitary egg is kept on the feet, which insulates it from the ice, and is covered by a special fold of the abdomen.

This display, so ceremonious and so grotesque to human eyes, is repeated on many occasions, usually in a sexual context. Fights may sometimes take place, in spite of the emperor penguin's famous placidity, and then the birds' attitudes are by no means the same. They face each other and try to jostle each other by thrusts with the breast, peck with the beak and strike each other with the wings. Sometimes fights get fierce enough for the two adversaries to become covered with blood, which seems greatly to alarm the peaceful groups of congeners in the area! More usually the "fights" are restricted to threats: the beak is pointed at the adversary with a growl of varying intensity; if the threat is meant to be more serious the beak and wing are raised as though ready to strike.

Generally all goes well and penguins walk around, alone or in groups, at a stately pace of from four to six kilometres an hour. But if they are frightened, for instance by the presence in the vicinity of the terrible killer whale, they use a much quicker method of locomotion: this is the "toboggan", achieved by lying on the stomach and sliding over the ice, rapidly propelled by the hind limbs; the speed reached is so great that even a man on horseback takes a good many metres to catch up with them.

But the colony is not silent; on the contrary penguins are rather noisy. First of all they have their famous courting song, a kind of chatter ending on a longer note with the males than with the females. There appear to be a large number of individual variations which, no doubt, are used for the purposes of recognition, as we have seen above. We may also note the trumpet call, a summoning cry so strong that it has been likened to a klaxon, that can be heard at more than a kilometre's distance: there are other cries too, such as that of fright, not so loud and deeper; also a cry of anger and one of satisfaction, a sort of "con-con" frequently made by the birds when bathing. The significance and importance of these sounds is clearly shown by an interesting experiment made by Prévost: he put a hood over a penguin's eyes leaving the ears free; although the bird was far from its congeners, as soon as it was freed it went straight towards them. But if the ear orifices. were carefully blocked and the eyes left uncovered the bird became bewildered, turned round and round and could not find the colony. This leads one to think that sound signals must be more important than visual ones.

SOCIAL ACTIVITY DURING THE REPRODUCTIVE CYCLE

Let us start at the beginning, that is to say towards the end of the "summer", mid-March in the Antarctic. The icebank has then been broken up for some time into isolated floating floes on which the penguins establish themselves; they travel long distances from their breeding grounds, but nevertheless find these last again without mistake by some as yet unknown mechanism. Remember that penguins travel on foot and by swimming, and not by flying, as pigeons do, so the penguins' homing is yet more remarkable. The sea starts to freeze over again in March and this is the time the first penguins return, usually in Indian file. The ritual of the "return" is always the same and it has been dramatically described by Prévost. "A few metres from the budding rookery the new arrival raises its head, stretches out the neck, then, invariably ... rubs the upper parts of the sides of the head on the right and left, once, twice or several times with its wing. It may well be that this very precise movement, made by all the newcomers, is done with the object of unblocking the birds' external auditory canals. . . . However, the fact that this is done by almost every one of the new arrivals might equally well lead one to suppose that it has some other purpose, at present unknown to us. The bird then gently bows its head towards the ground, and at the same time draws in a deep breath: it then sings, the beak held vertically downwards, then slowly lifts it up and listens. After a pause, which can be of varying length, the penguin mixes in with the main body of its congeners and, continuing its song, moves around the rookery."

I mentioned the return home: do couples really find each other? This question has been the subject of some dispute; it is now, I think, settled by Prévost's observations and by ringings showing that reunion is the rule and "divorce" the exception. For instance, as we shall see in a moment, after the male has "sat" on the egg for a long time the female looks for him all over the colony in order to take up the duties of incubation in her turn.

But couples do not always come together without difficulties: for instance, two females may both want the same male; "trios" are then formed; trouble is in store and quarrels between females can occur, with the male usually remaining passive. This can last two or three days, particularly at the start of the breeding season. Then the pair is formed and scarcely ever separates, even when they press together in the "tortoise" in bad weather; the two birds stay side by side as much as possible, one head leaning towards the other; sometimes the male may lie on his stomach and the female will slip her head under that of the male. Then sexual excitement grows; the male, the more active partner, inclines his head and dilates his incubation pouch, an action which the female soon imitates. The male rests his head on the lower abdomen of the female. At the end of the sexual display the female falls on the ice, spreading her wings and legs wide. The male mounts on her with considerable awkwardness; he seems to find it difficult to keep his balance. His actions appear to interest the other males, who draw near to watch more closely and try to jostle the male in order to capture his mate. The egg is laid in May–June: it is single, weighs nearly a pound and is laid with great difficulty; the female appears to suffer a great deal, and the male seems upset and runs round her; sometimes the female circles the male, not sparing him pecks from her beak, which he seems contritely to accept. As soon as the egg is laid the female puts it on her feet with the help of her beak; the feet insulate the egg from the ice. The male then announces the success of the operation by singing, accompanied sometimes by his partner. But he also is very anxious to incubate the egg himself in due course. His behaviour is most amusing and has been so vividly described by Prévost that I can not do better than quote him again. "The male inclines his head, and points his beak at his partner's incubation pouch: the partner shows him the egg, inclines her head as well and sings, in which she is soon imitated by the male. . . . After that the male frequently looks at the egg, at times touching it with his beak whilst softly calling, the body trembling the while. He draws in his abdomen making the incubation pouch more and more apparent. He progressively adopts the incubating stance, his body then resting on the posterior plantar cushion. Sometimes the impatient male inclines his head towards the female and even pushes her, trying to get hold of the egg by force. His partner then slowly separates her legs and soon the egg falls on to the ice. . . . Shaking with excitement, his tail going up and down, the male uses his beak to roll the egg between his own feet, on to which he eventually gets the egg with considerable difficulty. The couple sing and the female starts to walk around her mate who then becomes calm and indifferent. . . . The female moves off, comes back, sings and displays again. This coming and going is repeated and the female moves a little further off each time, swaying the body to left and right as she goes. She turns round the male, lifting her feet much higher than in normal walking, more balancing to left and right follows . . . then she goes off on her foraging trip."

It must be borne in mind that such a journey is very necessary, for the females fast from the time they join the rookery until after the egg is laid. They do not come back for two months, during which time the males continue to incubate the eggs, and to fast.

Males do not recognize their own eggs and if offered several choose the nearest, but there is very little chance of them losing their egg and there is no common sitting as some authors have maintained. There are a few unoccupied males that get relatively more excited than the others: these are the ones who try to get hold of a female when she is ready to mate, and when sitting time comes along they seem to be very interested in the eggs. But they have never been seen to capture one.

By the beginning of July the sitting males have lost a great deal of weight and their plumage is dull and often soiled with excreta; they never break their fast. It is at this time that the females, fat as butter, come out of the water and take over the sitting from the males. As I have already mentioned, they look for their own male and egg, singing as they search, and we must note that each one finds her mate among several thousand birds. They hesitate a bit at times: for instance a male, drawn by the song, may approach a female and start his display, but the female quickly leaves him and starts looking elsewhere. When a reunion is successfully achieved the male dives off in his turn to satisfy an appetite which he has had plenty of time to acquire. It sometimes happens that a female cannot find her mate, no doubt because the egg he was incubating has died; in such cases the male throws the egg away and goes off to fish. Unattached females are very desirous of acquiring penguin chicks and try to steal then from other females.

Some chicks hatch in the male incubation pouches, but the majority are born after the females' return: their raising starts in July and finishes in December, when the rookery breaks up. Until the beginning of September the young stay in the pouch where they are nourished by the regurgitation of food from the mother's stomach. Mothers easily recognize their own young; this one can verify by giving a hen bird two, her own and a stranger's: she will make no mistake and will drive the other bird's chick away by pecking at it. The bird recognizes the chick from the noise it makes: recordings show that no two are alike. By the beginning of September the young are too big to get into the pouch and suddenly leave their parents; they learn, for better or for worse, to get together in the "tortoise" when the weather gets bad. But loose groups of young birds remain when the temperature becomes more clement: these are the famous "nurseries" of the English authors. Prévost points out that this is not a good term, for no one looks after these young birds or takes any notice of them—though care of the young is found in other species of birds. In spite of this, young birds quite readily follow adults.

At the first moult, the young bird, which up to this point moves about quickly, becomes very fat and acquires the dignified appearance of the adult. This is the austral spring; the ice breaks up into big floes, which carry off adults and young, and it is on these floes that the young grow up.

CHAPTER SEVEN

THE PLACID HERDS OF THE BIG MAMMALS

COWS

OBVIOUSLY ONE would not expect to find complex social phenomena among domestic animals. Nevertheless they do exist, complicated by man's continual interference and the difficulties of domestication. Heifers, studied by Schein and Fohrman in the United States, do not have the calm and placid nature generally attributed to them. In the same way that there is an etiquette of behaviour among gorillas (see below) so also is there a complete order of precedence among cows, one reached and maintained by butting with the horns.

The process by which this order of precedence is established can be split up into the *passive approach*, which is much used as the animals walk long distances and have every opportunity of crossing each other's paths, and the *active approach*: this can be seen when a new heifer is introduced into a well-established herd. Then some of the animals move towards the new heifer, obviously with aggressive intentions, shown by a forceful

Fig. 55. Attitude of musk-oxen in the defence position against wolves. (After Zimmerman in Grassé's *Traité de Zoologie*)

blowing, lowering of the head, slower movement than is usual and scratching the ground with a foot. The next stage is *the threat*, which takes place about a metre away from the new arrival: the dominant animal, head down, faces the new animal and seems to want to run her through with her horns. If the new creature is aggressive, she also will lower her horns in the same way and a struggle will develop; if she is willing to give way she rapidly withdraws to a considerable distance.

A fight may start in some cases; the animals breathe fiercely and circle round each other; at some point this circling stops and the attacker tries to pierce the flank of her opponent with her horns; the latter naturally attempts to parry the blow. If she fails she will immediately run away; but it often happens that she succeeds in turning in such a way as to present her head and not her flank. If the attacker flees the victor will pursue for a varying length of time. If two cows are of nearly the same rank, the fight may be renewed at intrvals of a few seconds or minutes: between fights the two antagonists may put out their tongues and grab a mouthful of grass, but they watch each other out of the corners of their eyes. When the fight has lasted a while the less aggressive animal will tire, and allow her adversary to slide towards her flank, but at the same time she will put her head between the hind leg and the udder of the rival, which seems immediately to stop the attack. In short, as the attacker is about to strike she feels the nose of her victim against her teats and quickly stops the attack.

Social organization among cows is very simple. Dominance depends, above all, on age and weight. When a strange cow is introduced into a community she is smelt and even threatened by the other cows, but she in no way disturbs the established "peck-order"; she is merely put at the bottom of the ladder, even if her original rank and weight were reasonably high and her age reasonably advanced. Two things can upset the social mechanism; first of all oestrus. When a heifer is on heat she will sniff even the dominant beasts, especially in the vaginal region, and will repeat the attempts several times in spite of rebuffs, trying at times to mount on their backs. Secondly, when a cow has a calf she will spend a considerable time licking it, and no

Fig. 56. Dominating and fighting attitudes in zebras (*Equis quagga*), after Backhaus.

a. The animal on the right, raising its head, is probably taking a dominating attitude.

b. Threat.

c. Bites on the foreleg and back.

d. The animal on the left is trying to push its adversary's head towards the ground to stop it biting.

e. The animal on the right has adopted the attitude of "humility", at the same time striking the ground with its hind hoof.

f and f'. Biting.

f". Mutual exhaustion.

g. Kicking fight.

h. The animal on the right has just given up the fight.

i. The victor puts its head on the rump of the vanquished.

longer forms part of the herd, which she does not particularly follow as it moves around. When the calf is taken away she spends three very troubled days, mooing constantly; she then returns to the herd and is reintegrated into it.

Integration is not too strong a word here because cows

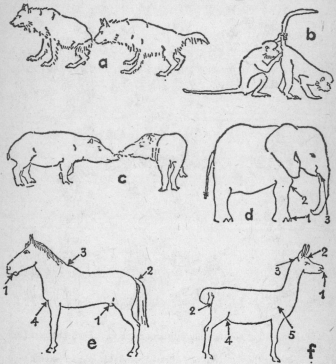

Fig. 57. Recognition procedures in some animals. (After Schloeth)

a. Naso-anal type.

b. Naso-genital type (*Macaca irus*).

c. Naso-nasal type (tapir).

d. Places sniffed in succession when two elephants meet.

e. Points successively sniffed at in the horse; the numbers indicate the frequency (in descending order) of the places smelt, figure 1 is the place most frequently smelt.

f. Order in which places are sniffed in the llama.

conduct themselves, if I may say so, rather like Panurge's sheep —that is, they do everything together: they all feed at the same time, they all ruminate together; but if a simple wire fence separates the herd into two parts each then becomes independent and one section may ruminate whilst the other crops the grass. The desire for grass cannot be very insistent except in very undernourished cows, for if one goes to the drinking trough the others will follow.

A DIGRESSION ON THE MYSTERY OF DOMESTIC ANIMALS

I am well aware that what I have written does not amount to much; you might think that there was not much more to be said on cows: but you would be wrong (as we shall see later

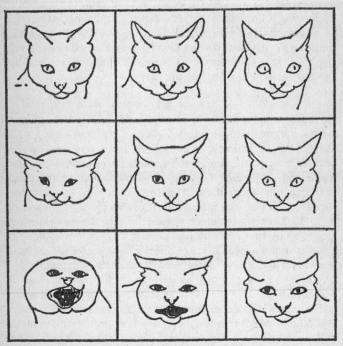

Fig. 58. Facial expressions in the cat. (After Leyhausen)

when we discuss the wild cattle of the Camargue). In fact, strange as it may seem, domestic animals have hardly ever been the subject of scientific behaviour studies.

Think of dogs and cats; man so projects himself into these animals that it is almost impossible for him to view them objectively; but if it could be done, if someone could find a method of doing it, no doubt we should soon find that the behaviour of the cat and dog is but partly known to us. As to our flocks and herds, we are too used to seeing them, and we no longer know how to discover in them the strange and the unknown: it is an axiom that cows are too stupid to be studied. As to sheep, the last word has been said when we quote the case of Panurge's sheep: nevertheless social integration is pushed to such lengths in a flock of sheep that they could be a mine of fascinating information for ethologists. And what about the great sensitivity of pigs, especially of sows, to the olfactory and auditory signals made by the boar? Do we know if these signals can accelerate the arrival of puberty, which is the case in other mammals, such as mice (see above)?

THE BULLS OF THE CAMARGUE

Be that as it may, the herds of cattle, half or completely wild, of the Camargue do throw a little light on the obscure story of the domestic cow. Schloeth, on horseback, has recently followed a herd for a very long time, some 2,000 hours of observation, and has noted some characteristic traits of behaviour (certainly found in our cows, though less markedly so). For instance, expression is not seen in the face, because to us the ungulates seem to have no facial expression at all; the emotions are expressed by the head and horns being more or less inclined in relation to the neck. These are direct indications of the animals' mood and are at once recognized by the rest of the herd; and any human observers with experience get to know them easily too. Smells are also of great importance among the methods of social expression: males often sniff the hindquarters of females, but seldom of other males, and it is also rare for a female to sniff a male. Then the mooing and lowing of the cattle are not as simple as one might imagine; Schloeth distinguished eleven different tones, among them the threatening bellow on the approach of a rival, the call which summons the herd together, the noise made by a cow to call her calf (which is very like that made by a domestic cow), and several ways in which the cow can "talk to her calf", as for instance the special call that draws its attention to something worth while. Some movements are more difficult to interpret: for instance, scraping the soil with a forefoot, rubbing the neck and the side of the head on the ground, always in the same place so that the grass gets quite worn away; rubbing things, such as the earth or, above all, trunks of trees, with the horns (and it is always the same trees

Fig. 59. Positions of the head and neck in the Camargue wild bull (after Schloeth). Note how slightly the attitudes differ when compared with those of birds, nevertheless they are readily perceived by the animals themselves.
a. Normal position.
b. Threat to the flank.
c. Position of approach.
d. Flight attitude.
e. Position indicating assurance.

that get roughly treated in this way). Often these gestures seem to be threats made in the presence of a rival, but they also occur when an animal is alone.

There is another group of social contacts which are very frequent but appear to have no precise relationship to the animals' rank; these also are difficult to intepret: among such is the licking of shoulders, often lasting a long time (most of the time the dominated animal licks the dominator), or certain contacts of the forehead and horns, which seem to be a kind of game, and one which does not develop into a fight.

For bulls do play, particularly in their youth, but the adults are not averse to taking part in games between two or more animals. There is a "call to play": a special cry which accompanies the twisting of the calves as they chase their tails, which they do almost as well as kittens. There are also "playgrounds" to be found—bare patches well separated from the rest, where the calves chase each other, play at fighting and rub themselves against the ground or bushes. Schloeth notes the curious fact that simply arriving in these special "playgrounds" releases the playing reaction, because before getting to the area the calves tend to be more or less quiet. They also race each other, usually in small bands which can include sub-adults, adults and even a large part of the herd. Finally they play a number of games with more or less marked sexual characteristics, such as pretended mating, or "mother and child" games with one calf pretending to suckle another. It should be noted that the habit of licking shoulders, so frequent among adults, is not found among the calves' games. Calves also play quite early with special objects, such as bushes, which they push with their horns or breasts; it is quite exceptional to see them take any interest in movable objects, such as stones.

The hierarchy is fairly complicated in the Camargue herds and, if one believes Schloeth, is subject to frequent change. As noted above, social contacts such as shoulder-licking do not have a very close relationship with dominance; nevertheless individuals ask for such attention from others within three rankings above them or three rankings below, never at any greater distance. This concept of "three ranks distance" seems to be a fundamental one and animals above or below these limits are treated as if they did not exist. It is within these three rankings that the positions are carefully guarded: it is to individuals within this distance that threats are made, leading sometimes to chases which imply a fall in rank for the pursued animal.

HERDS OF DEER

Fraser Darling studied deer in the Highlands. These creatures move about very quickly and have two territories; the high land for the summer and the low land for the winter. The winter territory is often bounded on three sides by natural features, such as cliffs or a river. In this area the hinds give birth to the fawns at the beginning of June; they stay there until the attacks of gadflies drive them to the high land. Males regard the area more as a refuge in time of storm and are less attached to it. The summer area consists of rounded hills running up to 1,000 metres above sea-level; it is a more restricted area and this must be the reason why males and females possess it in common, and do not live apart in separate herds as is usual. The most curious thing is the road network joining these two territories, which always follows the easiest route, a reason why men as well as deer use it. These roads seem to be very well known by the herd, which often uses them to avoid bad weather. Moreover this is no isolated case. Hediger has frequently emphasized that *the movements of wild animals often have a ritual character*, when they always follow the same path for the purposes of going to feed, bathe or defecate. In equatorial Africa many of the roads joining native villages are no more than rhinoceros paths, travelled regularly by these enormous mammals. The natives have been using them for a very long time.

The deer's society is different from those we have examined up to the present; *it is matriarchal*, and the males, however generously endowed with antlers, do not take over the direction of the herd, even at rutting time. Moreover the female herd is particularly well organized around an old hind; all females guard their young carefully until the third year. The chief hind always has a young deer with her, and when she loses her reproductive powers she also loses her position. She is very vigilant and watches for all possible dangers. Fraser Darling reports that one winter he put out food for a small herd of five animals with a leader hind, some young deer and another hind with her fawn. This small herd lived near his house in a very confined territory; he could easily see them and they could certainly see him. However, in spite of the food he put out during the whole winter, the herd remained as wild as ever, particularly the females, for the young males did become a little tamer although they never allowed him to get closer than 100 metres. One night, however, after he had put out the hay, he hid in the undergrowth. The leading hind came up soon afterwards with her little group, but seemed to be ill at ease; no doubt because she had not seen the man take his departure, as he usually did. She stopped five metres from the hay, her muzzle raised, whilst the others began to eat. She started to circle round, taking tiny steps, evidently a sign of distrust; then she saw the top of Fraser's head, which was sticking out; she drew a little nearer in order to see it better and then gave the rallying call. The other four animals raised their heads and came up to her, looking in the direction of Fraser, who had by this time sat down. As nobody moved Fraser went away through the trees to his house; only then did he see the chief hind begin to eat.

The young males leave the herd of their own free will at about three years old; they are not driven out. The stag herd is much less well organized than that of the hinds and has no apparent leader; moreover they may leave one herd without necessarily joining another, especially as they are more mobile than the females and move about much more. When resting and grazing they are often grouped according to age.

At rutting time the stags journey to the hinds' area. The first one to get there takes the entire hind herd as his harem, but when the other stags come up a redistribution takes place on a more or less stable basis of ten hinds to one stag. If danger threatens the hinds quickly reform their herd and take flight behind the chief hind, without appearing to worry about their males; the males may follow the hinds or go off on their own.

A matriarchal organization is also found among sheep: an old ewe leads the flock and is responsible for its safety. The rams are much more timid!

Mating lacks the strange habits adopted by half-wild herds of goats, or the impala antelopes studied by Verhulst. The males circle round the females several times and often for so long that a big circle is marked in the grass: this is what the English call "fairy rings". The same behaviour has been noted

Fig. 60. Interspecific relationships in fish (after Eibl-Eibesfeldt). The trumpet-fish (*Aulostomus maculatus*) regularly hides itself amidst a shoal of quite different fish (*Zebrasoma flavescens*), following all their movements; this allows it to approach close to its prey unperceived; it then suddenly darts like an arrow from the midst of the shoal of *Zebrasoma*.

in the American bison, but in a quite different context: the big bulls ceaselessly circle round the herd, particularly round the young, to protect them from coyotes. Among mammals some males pay as much attention to caring for the young as the females (for example, the hare) and others do nothing at all for them (the rabbit).

As to the relationship between mother and offspring, obviously they are very close in mammals, as the young are fed exclusively on milk. There is a little-known exception, that of the famous "eucalyptus bear" (which is not a bear at all), the koala. This mammal is a marsupial, having the ventral passage opening not forwards but backwards. As the time of birth draws near the female excretes a special substance for about a month beforehand; it is made of a mixture of eucalyptus leaves rich in peptones: as soon as it leaves the maternal rectum the young koala puts its head out of the marsupial pouch and at once eats some of this transition food.

INTERSPECIFIC RELATIONSHIPS

We know that these exist between mammals and birds. Buffaloes tolerate on their backs a winged flock which frees them from vermin. But what happens if, as in some zoos, two species are put together that have never met before, such as an African ibis and a big American rodent, the water-cavey (*Hydrochoerus capybara*)? Hediger was surprised to see the ibis determinedly jump on to the water-cavey's back and start to clean it of pests. The most striking thing was that it was done in the twinkling of an eye (*augenblicklich*) with no difficulty on either hand! This presents quite a problem to the scientist, for these reciprocal "acclimatizations" seem easy and frequent; the ibis is no exception.

CHAPTER EIGHT

TAKASAKIYAMA, OR THE FORMAL WORLD OF THE PRIMATES

FRISCH, WHO is well known for his work on the biology of primates and, in addition, speaks Japanese, two specialities rarely combined in one biologist, was one day attending a conference at the University of Chicago on the behaviour of primates. He noticed on a shelf in the library a number of works and a review, all in Japanese, covered with a layer of dust; it was obvious that no one had consulted these publications since the day they came. Though American ignorance of all foreign languages may be inexcusable, perhaps in the case of Japanese one can admit extenuating circumstances! Frisch, however, as we said, understood the language and he began to turn over the leaves. His amazement knew no bounds: he found himself reading an account of the profoundest and most intelligent observations on the subject he had ever seen, carried out by a whole team of Japanese who had spent years on the project, of which, however, nothing was known in the western world. And the subject was—the behaviour of a species of Japanese monkey! Frisch lost no time in telling the news to the conference and its chairman, who had had no idea that they possessed such a treasure in their library. The sensation it caused was immense, and completely justified, as we shall see, by the report which Frisch later edited.

Soon after the war Miyadi and Iwanishi of Kyoto decided to set up, in co-operation with another group who were interested in monkeys from the medical angle, a combined study of the biology of monkeys. It was chiefly concerned with some macaque monkeys (*Macaca fuscata*) which are fairly widespread in the southern islands of Japan, near Honshu on the north-east coast of Kyushu. The first group of monkeys lived on a small mountain, Takasakiyama, cut off on three sides by the sea, and on the fourth by highish mountain ridges. Never in the memory of man had the macaques been seen to leave this restricted habitat. The Japanese research team installed themselves near a little Buddhist temple beside which lay a field of sweet potatoes that attracted the monkeys: a convenient observation post. The Japanese adopted the methods perfected by the great Carpenter, which are practically identical with those used by Konrad Lorenz when observing geese and ducks. First one must learn to know the animals in the group individually, giving them names as soon as possible: in the case of these monkeys identification was made easy by peculiarities of coat colouring, which varies a good deal in this species. After that the workers divided the day into observation periods during which everything that occurred was noted down in writing or on a portable tape-recorder. However, the unusual thing about this study is that it continued for more than eight years, and the number of documents that accumulated in this time was enormous. One can only compare it with Lorenz' twelve-year-long observations on a certain grey goose. Obviously, I shall be able to give only a brief summary here, following Frisch as closely as possible.

To begin with, there exists here, as in all animal groups, a social structure, manifested in the field by a *concentric distribution*. The central part is occupied almost entirely by the females with their young of both sexes, and sometimes by a few big males. There were sixteen males at Takasakiyama, but only six had the right to go to the centre, and these were the biggest and apparently the strongest. The other males and those who were not yet quite adult were only allowed to stay on the fringe, in the trees or on rocks. Even there the distribution was not free for all: the "sub-adults" were banished to the extreme periphery, while the adults were allowed nearer the middle. The very young monkeys, however, were permitted to roam wherever they liked, and did not fail to do so. This is very reminiscent of Tinbergen's observations of Eskimo dogs in Greenland. These dogs, you remember, can behave like real wild beasts to each other and fight a trespassing member to the death; but if he is young, the punishment will be less severe.

This concentric distribution remains unchanged all day, and

the animals feed in their places according to their rank. But when evening comes they retire to sleep and then a ceremonial procession is formed, unvarying in its order: first go the dominant males with some females and their young. Only afterwards, when they are quite sure that all the "lords" have departed, do the second-class males enter the "sacred zone" of the group bringing with them the females and young that still remain, and taking over the duties just abandoned by their chiefs; that is to say they protect the group from intrusion by possible enemies, keep discipline, especially as regards pacifying quarrels, and give the signal for departure. Soon there are only a few stragglers left and it is now that the sub-adults dare to penetrate the centre in their turn. The few remaining adult males allow it and accept their help in rounding up the last females present. After that some of the young male sub-adults enjoy a romp on the territory before retiring, and finally a number of solitary males (there were three of them at Takasaki-yama) enter the territory, which they have not dared approach during the day, in the hope of gleaning some food scraps. And at dawn on the following day the procession returns in the same order to reoccupy its concentric zones.

When, on the other hand, the monkeys are not going to bed, but to forage in some new place, they still advance in procession, but this time the order is different: the vanguard consists of medium-sized males; in the second rank come the big males, females, some carrying babies, and a few walking young; the third is made up entirely of young monkeys, and another group of medium-sized adult males brings up the rear. Imanishi points out that this is still the same concentric distribution adapted to the form of a straight line.

During the day there is strict police control. The big dominant males watch over the females and their little ones in the centre, and prevent males of inferior rank from encroaching on the zone. Yet these lower males are allowed to co-operate with their chiefs in policing the group, chasing and biting law-breakers. They also co-operate with the sub-adults on the periphery, but in this case against enemies from outside. The sub-adults join in the guard-duty, but without unduly exerting themselves. They spend much of their time in play—rough play that may have a part in deciding their future social rank.

This rank, once established, seems to be maintained by a ceremony often found among monkeys: the subordinate one assumes the posture of the female and the dominant one makes the gesture of copulating with it. This occurs principally in the morning at the moment when the concentric order is being re-established. But it is only a show, there is no introduction of the penis: and when the subordinates present their hind-quarters to their lords in the morning it is all over so quickly that it seems like a mere salutation, as the Japanese research-workers say!

Rank is also expressed in the reactions to strange forms of food. For example, it was not always possible to keep visitors away from Takasakiyama or to prevent them from throwing caramels to the monkeys. Now monkeys at a zoo are quite used to these sweets and know how to take off the wrappings, but the Takasakiyama monkeys had never seen them before. With them unknown foods were not thought fit for the lords, and only the little ones picked them up. Later on their mothers took some, and then the lords who looked after the one-year-olds when the female was about to give birth again. But the last of all to get used to the caramels were the sub-adults on the outer fringe, who had no contact with the innermost circle. The pro-cess of familiarization (acculturation as the Japanese, rather stretching the anthropological vocabulary, call it) is extremely slow: it took nearly three years for the sub-adults to become accustomed to the caramels!

Obviously it is of primary importance to know whether the sample from Takasakiyama is typical of the rest. Do other groups of monkeys behave in the same way? In fact, they do not. They appear to possess a "culture" with different "traditions". The Takasakiyama tribe, for instance, seems to be very rigid and closed compared to the twenty other groups of monkeys studied in different parts of the country by Japanese biologists. They have much less individual liberty. In other tribes, such as that of Minootani, the sub-adults may form veritable gangs, which go on expeditions together lasting several days, and when food is thrown to the group they all gather round in a joyous hubbub without distinction of rank. Furthermore, the Spartans of Takasakiyama punish their wives harshly; sometimes all the females bear tooth-marks. The Athenians of Minootani have gentler customs and one hardly ever sees a scarred female among them. With the latter, social rank is maintained by a mere feint attack by the dominant male, whereas with the former the attack is often real to the point of biting. The same goes for the "acculturation" (!) to caramels: our Athenians became quite used to them in two months; our Spartans took three years. In both tribes the females are rather fickle, but the Minootani lord appears to regard with a paternal eye the gambols of his mate with a male of inferior rank; he attacks neither of them. But woe betide the Takasakiyama females who indulge in such sports; they are promptly and unmercifully beaten; as to their partner in sin, the lord has only to look him in the eye and he will flee without asking for any more.

This variability is typical of all their behaviour patterns; for example, in their search for food. In Takasakiyama they always radiate round a centre; at Arashiyama the territory changes with the season; at Shodoshima the tribe follows a sort of zig-zag. Perhaps the pattern of their peregrinations is handed down from generation to generation. There are also differences in their eating habits: some groups do not know how to eat eggs, while others gobble them greedily. In one group only adults over twenty years old eat them.

It appears, too, that small but numerous differences of vocalization embroider the broad canvas of cries common to the whole species. Research in this field has not yet advanced very far. In any case, it may be that such variations establish themselves and increase, in this domain as in others (among birds, for instance; see above), because the tribes have little chance of meeting and seem even to avoid each other.

Can it be shown that the group evolves with the passing of time? Here, too, the Japanese investigations were carried on long enough to make an answer possible, at least as regards the Takasakiyama tribe. It has grown from 160 individuals in 1952 to nearly 600 in 1958. And yet the number of dominant males has fallen from 6 to 4, all aged over twenty; and there are still only 10 of the "second rank". Guard duties have become less important for the sub-adults now that they are so numerous. Moreover, some of them have simply left the tribe. There have also been some attempts by the second-rank males to join the dominant caste, but with no success.

THEORETICAL MATTERS

Obviously all these facts bring to mind the ways of savage human tribes. That raises the question of whether, as with

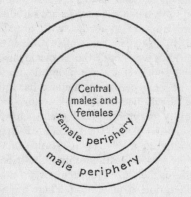

men, the family, in some shape or other, forms the hub of society.

Zuckerman, the famous primatologist, based everything on the family group formed by a dominant male and his harem, without any sexual promiscuity. However, certain observers have shown that there are monkeys who live in small bands with several males, where sexual promiscuity is very frequent, so that Imanishi asked what the word "family" really denotes if sexual promiscuity is the rule. He is rather in favour of replacing the word "family" by the Greek word "*oïkia*", which means house, the smallest social unit possible, whatever its composition. One can then distinguish two types: the *oïkias* with their own territory, hostile to other neighbouring *oïkias*, such as those of the howlers, the rhesus and the gibbons; and secondly, the *oïkias* that associate with their neighbours in order to form a larger band, such as those of baboons.

Up to now we have considered dominance only among the males, but it also exists among the females. At Takasakiyama, where all the phenomena are particularly distinct, the female class system runs parallel to that of the males, and the outer circle is dominated by the centre.

According to Kawai, however, two kinds of rank are to be distinguished in dominance: basic rank and dependent rank. Basic rank can be ascertained when two monkeys are alone together; dependent rank is shown in a situation where several individuals are together. The first to manifest itself is dependent rank, seen when the baby monkey is with its mother: it shares her rank, especially since it has no basic rank of its own. That it will acquire later, when it has become independent of its mother, in the course of games and squabbles with its young companions during the process which Kawamura, with a certain lack of euphony, calls "peripherization"(!). When the young leave their mother they also leave the central zone and lose their dependent rank. The same applies to the young females of Takasakiyama, but not to those of Koshima, who stay in the central zone, keeping their dependent rank until, as Imanishi says, it "solidifies into basic rank". This is even more in evidence in the Minootani colony, where there are no dominant males and where a matriarchal society seems to be coming into being. According to Kawamura two important principles determine rank at Minoo: the first is that the child's rank corresponds to that of the mother, and the second that among siblings the younger ranks higher than the elder. Imanishi adds that, as we have already seen, the apprenticeship period and reactions to different situations depend on rank. Children of dominant females automatically learn dominant attitudes, those of subordinate females the attitudes of submission. The children of the dominant females in the central zone are also nearer the lords than those of the subordinate females in the periphery. They thus tend, says Imanishi, to "identify themselves more easily with the leaders, to be accepted by them more readily and *finally to succeed them*" (the

italics are mine). Yamada claims to have observed a case of the
succession of a young male.

The portrait of macaque life painted by the school of
Imanishi is a striking one. As you will have already noticed, it
overlaps in more than one respect what we have learnt of the
life of primitive peoples. And, for all we know, it may go further.
I am thinking of the concept of taboo, for instance. Are there
forbidden objects or places among monkeys? Some data rather
suggests it.

Anthropologists have hotly debated Imanishi's theories,
especially his tendency to designate certain of the animal's
characteristics as "sub-human". I grant that he may exaggerate
on occasion and that some of his analogies may be superficial,
but for goodness' sake let him get on with it! Let us see where
their special cast of mind, felicitously assisted by an impeccable
technique, will lead us. He has, at all events, as everyone agrees,
opened new horizons in primatological research.

THE BABOONS OF AMBOSELI

Astonishing though the macaques may be, it seems, in the
light of recent research, that baboons are several lengths ahead
of them. They are typical social ground monkeys with the habit
of attacking intruders in a fairly well co-ordinated formation,
all the more dangerous in that the males are very strong, with
formidable tusks, and are known on occasion to be carnivorous.
There are biologists who claim that in some respects the baboon
is more advanced than the chimpanzee.

Washburn and de Vore were able to observe them from very
near in the nature reserve of Amboseli in Kenya. The actual
average of a tribe of baboons is 80 monkeys, but the numbers
may vary from 12 to 87. Each tribe has a territory of fifteen
square kilometres at the most, but it is a theoretical territory,
for only small parts of it are much frequented. Furthermore,
the frontiers are not strict (in contrast to those of macaques)
and Washburn once saw up to 400 baboons drinking from the
same stretch of water. . . . They consisted of three tribes temporarily together, but not mixing at all. All the same, such meetings appear to be rare.

When the tribe moves on it assumes the form of a procession, regularly organized like that of the macaques, but with a different order: first come the subordinate adult males with a few sub-adults, then the females with other sub-adults, then the dominant males followed by females with little ones and adolescents. The females carrying babies are the heart of the procession and after them the end follows in the same order as the vanguard, dominant males bringing up the rear. The males protect the colony very efficiently: they have only to adopt a threatening attitude for dogs or even cheetahs to turn tail and run. Lions are the only animals that make them afraid, and then they will take to the trees. Lions are also the only animals that will attack the fearsome tribe of baboons with any determination; the others seldom carry aggressive behaviour against them very far.

Baboons are very social animals, never found alone, far from their fellows. A wounded baboon, unable to follow the troop, is practically condemned to die. It would be interesting to know whether aid to the wounded exists. It does not seem to have been observed up to the present.

On the other hand, these monkeys often appear to have social relations with other species, especially with antelopes who, like all the ungulates, have a very keen sense of smell. The warning barks of the baboons, never far from the antelopes, alert them at once, and in the same way when the antelopes give the alarm the monkeys run away too. The antelopes scent danger, whereas the baboons perceive it through the eye: thus the two species co-operate, each by means of its most highly developed sense. The cheetah's favourite prey is the Impala antelope, but when these find themselves beside a tribe of baboons, they do not stampede but calmly watch the large baboons chasing the big cat away. In the mating season, when the male Impalas fight with their horns, the baboons will continue to eat, quite undisturbed.

These monkeys all spend the night high in the trees, an efficacious protection against carnivores and big snakes, which chiefly hunt at night. Baboons are specially nervous of the dark and do not come off their perches till it is broad daylight.

During the day baboons indulge in an activity that appears to us very unromantic, but which holds an important place in their lives, and that is *de-lousing*: individuals present themselves to each other to be de-loused. The de-louser separates the hairs with his hands and removes vermin and dirt with his mouth. During this operation the patient closes his eyes and gives every sign of perfect bliss, after which the two monkeys exchange roles. It is a complicated ceremony in which rank must not be forgotten. The centre of attraction is often a dominant male or a female with her baby. The lords of the tribe seem to be the most attractive and can hardly sit down for a moment's repose without a vassal hurrying to de-louse them.

Washburn notes with truth that the opposite obtains among herds of ungulates, such as the Impalas: here the big males spend their time trying to keep everyone together while many of their rebellious subjects show a strong tendency to stray. With baboons, on the contrary, the big males are so attractive to the others that they want to stay as near as possible to them all the time. The newborn are nearly as interesting. The chiefs hardly leave the young mother's side, either at rest or on the move. When she sits down the adult females and adolescents gather round her, de-louse her and try to do the same to her baby.

These English observers have also brought to light what appear to be friendships between monkeys. Some individuals are nearly always together, especially adult females. As to the young ones, they form little groups with their playmates that may last for years. As soon as the youngster begins to eat solid food and to leave his mother for longer and longer periods, he immediately joins one of these groups, in which he will probably serve his apprenticeship to social life. The games they play are essentially of a teasing kind and include scuffles ranging from playful to serious. Sometimes, just as with children, one of the more vigorous youngsters will go too far and provoke cries of pain from his companions. Then—as with us humans—an adult will come up, separate the combatants and distribute a few clouts, if necessary, and the game will stop. As many writers have pointed out, the arbitration of disputes seems to be one of the attributes of dominance, among birds as well as mammals; not only quarrels between young but also those between adults are stopped by the leaders at an early stage. With baboons it goes further: it would appear that the disputants deliberately fight or quarrel in front of the leader, who then "pronounces judgement" which, though limited to a few barely perceptible grunts or gestures, proves very effective, for the argument ceases forthwith. For my part, I am inclined to detect in this a phenomenon different in kind from dominance and still something of a mystery.

Dominance is also expressed when food is thrown to the baboons. It is the head of the hierarchy who comes to pick it up. The others do not even look at it; to so much as throw a glance in that direction would be a blunder of some magnitude, for among baboons to look someone in the eyes is a challenge that leads inevitably to combat. The same applies to gorillas, and it was because he understood that so well that Schaller was able to get himself accepted in one of their groups. If you approach humbly and, above all, look no one in the eye, they will not object to you.

One essential factor of dominance is that it works chiefly on a group level, for the adult males usually stay together and will all come to the rescue if one of their number is threatened (except, as we have seen, in the case of a wounded comrade). No doubt this is why the hierarchy remains stable, often throughout several years: a strong male, capable of beating any of the chiefs individually, cannot tackle them all at once. Basically the hierarchy helps in maintaining social stability, especially as one of the lords' functions is to keep order; serious quarrels seldom occur. Another still more important function is to defend the young and the weaker members of the tribe, and this may be one reason why these so often come to sit near the big males.

What about sexual habits? It seems that we must to some extent depart from the theories of Zuckerman, who saw in sexuality the mainspring of primate society. The females are receptive for only one week at most in the month. At the beginning of the oestrus the female will approach the young and subordinate males, who will mount her; but later, at the height of the oestrus, she "presents" herself to the lord. The male may not be interested, in which case she does not insist but notwithstanding does him the courtesy of de-lousing him. It will not be difficult for her to find another gallant, for chastity is not the baboon's strongest point. The two lovers will then act as a couple, but not for more than a few days. Battles for the possession of a female may occur, but they are rare and only take place when the hierarchy has not yet been definitely established. The males are monogamous for the brief period of their love affairs; the female in oestrus abandons her little one and goes from one male to another. It can be seen that, contrary to Zuckerman's generalizations, there is nothing among baboons that resembles a family or harem.

On its own territory the group feels at home, sleeps on special trees, drinks at one special place and eats at another. All this contributes to a solid integration of the individual, and it is extremely rare for a monkey to leave its tribe and join another one.

But we cannot leave this chapter on baboons until we have had a close look at the interesting studies of sentinels and protection behaviour undertaken by Dr Hall of Bristol.

BABOONS ON GUARD DUTY

Naturalists have always noted the presence of sentinels among baboons. Back in 1913, for example, Elliot wrote: "When they are engaged in a possibly dangerous expedition a sentinel is always to be found in a suitable spot for giving the alarm at an enemy's approach, or chasing predators." Allee in 1931 noted that "the guard is extremely alert and can detect the faintest sound, scent or glimpse of man or leopard. . . . The sentinels are often among the strongest males, excluding the leader of the tribe. . . . When their warning cry reaches the chief's ear he immediately assembles the other dominant males, arranges the males on the outside and the females with their young in front of, inside or behind the line of males. As to himself, he runs up and down at the head or rear of the group according to the plan of flight and degree of danger. When things become too hot for the sentinels they retreat a short distance, climb on to some high place and once more give their chief a warning cry."

All very fine, but Zuckerman does not agree about the part played by the guards; he even doubts whether they exist. He argues that it is possible for a big game hunter reaching the summit of a hill suddenly to find himself among a troop of baboons on the other side, which could not happen if there were sentinels. This is true and has been confirmed by more than one biologist. Who, then, shall we believe? Hall decided to get to the bottom of it and went to the Cape of Good Hope nature reserve, where he spent seventy-six days among baboons. It turned out that the question was more complicated than had been supposed. The big, watchful males noticed by so many observers do exist; but it is equally true that a man can land almost in the middle of a band of baboons without disturbing them unduly. Apparently it all depends on a complex set of circumstances which it is not easy for us to analyse.

Sometimes Hall saw isolated animals, usually young males ahead of the main body of the foraging tribe, give the first alert that observers were near. Females also sometimes take guard duty. But the true sentinels are the big males: they scan the surrounding country and sit till all the troop has passed by, moving only their heads, turning incessantly, the better to espy any possible ambush. What is extremely interesting is that if during this time females in heat come to provoke them they pay absolutely no attention to them. The big males may also be put on the alert by the barking of a younger one: they will then hurry to the spot to find out what is wrong. But only in very urgent situations will they themselves utter the double bark, the characteristic danger signal, to which the whole tribe reacts violently.

There are at least two or three situations when the behaviour of the guards can be observed at leisure: when the presence of men is detected by the baboons in the early morning, just as these are coming down from the tree-tops where they have spent the night; when the observers have already been noticed, but a sudden mist has hidden them from view; or lastly when a neighbouring tribe trespasses on their territory.

R

It is quite true that *except in these circumstances* a man may approach fairly near to the baboons without any very marked reactions on their part. The national park where Hall worked was crossed by several roads, so that the monkeys saw quite a number of visitors. One group, nevertheless, stopped short at the sight of observers, "posted its sentinels" and waited several hours before deciding on a detour despite the fact that they were probably hungry. In the same way the big males showed great vigilance before the troop crossed a road.

Hall divided the baboon's guarding behaviour into three kinds: (1) *Accessory vigilance,* or alarm given by an animal temporarily separated from the majority of the tribe; (2) *dominant vigilance,* practised by the big males, who may have been alerted by others, and who control the troop's behaviour in cases of danger; (3) general vigilance manifested by various cries from the young and females.

Let us add that these sentinels, so typical of the baboons, are not without their equivalents among other monkey species.

THE PEACEFUL WORLD OF THE GORILLAS

If Schaller and Emlen are to be believed, these enormous creatures have the gentlest, calmest manners in the world. One of these two observers grew to know their habits so intimately that he was accepted by a band of gorillas and could move about among them without causing any anxiety, even sleeping beside the big males. To succeed in this it is not only essential to give oneself up heart and soul to the project and to know the gorilla code of manners inside out: it is vital, above all, never to look an ape in the eye, for that is rank discourtesy, a gross threat for which you run the risk of having your head torn off by a careless swipe from a justly enraged gorilla. Schaller followed these apes in an almost unexplored part of the Virunga volcanoes in the Albert national park, and was able to observe them during 457 hours: at this stage individual recognition of the subjects was possible.

They are strictly forest-dwelling and vegetarian, humid forest being their favourite habitat, even though its aspect and vegetal composition may vary considerably, according to the altitude, for example. They are not climbers; their natural gait is on all fours on the ground and it is only exceptionally that they stand up on two feet. Instead of taking refuge in the trees when danger threatens, as baboons do, they climb down down if they chance to be there and take flight, running along the ground. A tribe of gorillas contains from 5 to 27 animals, the average being 16. There are, in the first place, one or more males with white hairs on their backs, showing that they are at least ten years old, some adult and sub adult females, one or more immature adult males and a varying number of adolescents and little ones. Isolated males have sometimes been found at least thirty kilometres from the nearest group, but this is rare and in general the tribe is close knit and feeds or rests (a little like the macaques of Imanishi) in a circle at least 60 metres in diameter. Occasionally the strength of the tribe varies as, for example, when strangers join it. Schaller notes one case where seven males with white hairs on their backs associated with a strange tribe during a period of twelve months. Thus the tribe is not entirely closed, owing no doubt to the gorilla's paternal character, of which we shall have more to say further on. Furthermore, these meetings between groups are quite peaceful and gorillas have no regular territory. Their moves may vary from 100 metres to 5 kilometres in a day. Inter-tribal encounters do not seem to cause any very definite reactions: either the members of the two groups forage side by side without mixing, or else there is a transfer of partners, mutual or otherwise. Schaller confirmed one such fusion that lasted for several months. He never saw actual battles: at the most only an exchange of threats.

Relations are just as peaceful within the tribe. To begin with, there seems to be no competition for the females, and food is usually so plentiful that there is no need to embitter one's character on that score either. The well-known ceremony of de-lousing, so habitual with baboons, is almost non-existent among gorillas. They must have some other way of removing dirt and vermin, for their coats are very clean; though it should be noted that baboons, probably because of the de-lousing, do not suffer from the ticks and other parasites that torment most mammals. Dominance is not entirely absent from the peaceful world of gorillas, but it could hardly be called draconian. It is all summed up in the rights of certain individuals to take precedence or sit in a special place, privileges apparently unchallenged by the other members of the tribe. Only once, on a question of feeding, did Schaller observe a slight tension in the atmosphere, and even then the dominant male confined himself to giving the upstart a tap on the back and a very short stare in the face. The latter immediately resumed his former respectful attitude and no further violence was necessary. The grizzle-backed males dominate all the other apes. The females and black-backed males dominate the young ones. One of the males is the leader: everyone watches him and copies what he does. If he builds himself a nest of branches, such as gorillas sit on to sleep, in a tree or on the ground, the others build one like it. If he moves on, the others follow.

As to sexual relations, these appear very insipid compared to the goings-on of the baboons and chimpanzees, and they are not easy to observe. The dominant males cast a paternal eye on a female copulating with a young male, and make no attempt to interfere.

The mother-child relationship is closer, in the nature of things. The baby is carried in its mother's arms during its first three months and does not try to play with others until it is five or six months old. At this point it begins to feed itself, milk becoming a mere secondary food. It is only between the infant and the females that de-lousing takes place, and it will lose its interest for him soon afterwards. Even when the mother has long ceased suckling him, and even if she has had another baby in the meantime, some degree of special relationship will be maintained between the two until well into the second year.

Communication within the tribe is almost entirely by means of gesture. The voice is little used: when two females are quarrelling, for instance, the male may give vent to a brief but piercing "Ooh, ooh", which will pacify them at once. If the same sound is made when all the tribe are eating peacefully, all eyes will turn to the leader, and then to the quarter that has attracted his attention.

Several kinds of apes make nests in trees, but the gorillas' nests of which I wrote are usually on the ground. They are little more than heaps of broken branches piled on top of one another, and are never used twice, for the simple reason that the gorilla defecates on them.

Anything that moves or seems alive rouses a gorilla's curiosity and they will even come right up to an observer who is alone— and not a nervous type! Objects such as a paper bag or a jam-pot, however, do not interest them. They have not the chimpanzee's passion for handling things. In the thick forest which is their home nobody dare attack them, and they have only to stretch out their hands to pick a succulent morsel, which explains their general apathy and lack of interest in using their hands. The chimpanzee's environment is more marginal, so that he has to make more effort.

Schaller and Emlen make the ingenious suggestion that the differences of motivation and behaviour observed in our groups of primates may well have been present in proto-humans: some may have been manipulators and lascivious like the chimpanzees, others rigidly regimented like the macaques, still others epicurean and placid like the gorillas. There is nothing to say that Australopithicus and Paranthropus shared the same temperament. . . . But there is one big difference, and that is that the attachment to a home, so marked among carnivores, rodents, fish, birds and men, is lacking in primates. And there are other particulars which it will be interesting to summarize in a table, after Washburn; I should like to complete it by adding some information concerning macaques and gorillas.

CONCLUSION

In the course of this lightning tour of animal societies we have had to move rather abruptly from one world to another. As we approach the conclusion of this work I should like to draw attention to the relationship of human society to animal society.

Let us begin with the relationship that once seemed so clear to us. Young and developing sciences, such as human sociology, often pass through successive periods of contradictory obsessions. The early sociologists, as we have seen, were fascinated

Table

Ecology	Baboons	Macaques	Gorillas	Primitive men at hunting and food-gathering stage
Size of group	10-200 per group	100-600	20	50-60
Territory	7-15 sq. km. No defence of territory.	20 or more sq. km. Territory defended to some extent.	Indeterminate. No defence of territory.	500-1,500 sq. km. Territory fiercely defended.
Population Structure	Small endogamous groups	Small endogamous groups.	Small endogamous groups.	Tribal organisation, with exogamous groups.
Diet	Vegetables; meat occasionally.	Vegetables	Strictly vegetarian.	Omnivorous; men specialize in hunting, women in food-gathering.
Economic dependence	Young independent as soon as weaned.	Young independent as soon as weaned.	Young independent as soon as weaned, but young have relations with mother even after weaning.	Child dependent on adults for years. Hunting, storage and sharing of food.
Social organisation	Closed, self-sufficient society; sub-groups based on age and preference, but temporary.	Closed, self-sufficient society; sub-groups based on age and preference, but temporary.	Closed, self-sufficient society; sub-groups based on age and preference, but temporary.	Inter-group affiliations, and dependence. Sub-groups based on blood relationship.

	Based on dominance and physical force.	Based on dominance and physical force.	Based on dominance and physical force.	Based on custom.
Social control				
Sexual behaviour	Oestrus among the females. Multiple unions, no family life.	Oestrus among the females. Multiple unions, no family life.	Sexual relations very fleeting and of little importance.	Female receptive all the year; family stable, at least for some years, often for life. Incest taboo.
Mother-child relationship	Intense but brief.	Intense but brief.	Prolonged during one or two years.	Very prolonged, the new-born baby being helpless.
Play	With other individuals; highly developed.	With other individuals; highly developed.	With other individuals; highly developed.	With other individuals, but also with inanimate objects.
Inter-communication	By gestures and numerous cries, but referring only to the situation of the moment.	By gestures and numerous cries, but referring only to the situation of the moment.	As Macaques, but cries very rare.	A language, which plays an essential part.

by the panorama of insect societies, about whose internal working little was then known. Nowadays we realize that to set up the ant as a model for man is, from a scientific point of view, irrational.

For the same reason we should beware of over-hastily comparing primate societies with those of man. They are still far behind the Stone Age. Even the most primitive group of Australian aborigines surpasses the rough and brutish world of baboons and macaques infinitely and in every way.

I agree that there are analogies. They must have caught the reader's attention more than once, and they go further than was dreamt of even ten years ago. It may be that bisexual, organic beings cannot come together without a certain number of basic mechanisms being set in motion, whether they be birds, baboons or men. However, data on animal sociology are not yet plentiful enough for us to be able to chart these analogies and differences with an accurate pen.

Nevertheless, there are two points at which comparison with man seems clearly justified: I refer to those of dominance and territory. Dominance among children is without doubt established in much the same way as among chickens; and which of us has not known *alpha* and *omega* schoolchildren? We must recognize these parallels, if only to protect *omega* from others, and *alpha* from himself. Who can say what distant echoes and complex results of ancient hierarchies are still at work in those savage creatures that are our children? As to territory, alas, I wonder if we are any more rational there than the grouse. Is our behaviour different from his in any respect? This question haunts me, and I incline towards a negative reply.

Let us at least hope that the segregation slowly dividing scientists of different disciplines will not prevent communication between child psychologists, ethnologists and animal sociologists. Frequent mutual discussion would certainly build up an image of society and pinpoint man's place in Nature in a way that would take us all by surprise—to the benefit of everyone.

And the insect societies? What fluttering there has been in the theoretical dovecotes, and how much better we could fit the pieces together without it! As I said at the beginning, the insects came near to winning the game. But I think that for our understanding of them this planet and our sciences are both too narrow. If only we could know what solutions life has adopted elsewhere in the universe: then we should command a wider, more intelligible context. Let us hope that new, as yet undreamt-of comparisons will open up for us; for to be able to compare is half way to understanding.

CONVERSION TABLES

TEMPERATURE

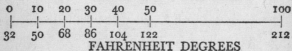

DISTANCE

kilometre (km) = $\frac{5}{8}$ mile

metre (m) = $1\frac{1}{10}$ yards

centimetre (cm) = $\frac{2}{5}$ inch

millimetre (mm) = $\frac{1}{25}$ inch

WEIGHT

kilogram (kg) = 2 lb

gram (gr) = $\frac{1}{1000}$ of kilogram

milligram (mg) = $\frac{1}{1000}$ of gram

The tables for Distance and Weight are approximations.

NOTES

NOTES ON PLATES

I. Development of the worker bee (R. H. Noailles).
Top left: The first day, the egg is placed vertically.
Top right: Larva on the fourth day; it rests in a curled-up position on the royal jelly.
Bottom left: Eighteenth day; the larva has become a pupa and the eye is becoming pigmented; the adult body is starting to develop.
Bottom right: Twenty-second day; the pupa just before eclosion; the wings are visible.
II. Collection of pollen by bees; a nectar forager covered with pollen (M. C. Noailles)
Bottom: Workers returning loaded with pollen, seen attached to the hind legs (Guy Dhuit).
III. Fight between queens; and bees fanning.
Top: Fight between two queen bees (Rapho, photograph by Frank W. Lane).
Bottom: Bees fanning; ventilation is effected by the rapid beating of the wings, so fast as to be almost invisible in the photograph (Guy Dhuit).
IV. Queen bee laying eggs and being licked by workers.
Top: Queen, surrounded by workers, laying eggs. The insect carries a metal disc attached to her thorax bearing the "colour of the year". The colour is changed yearly and the disc is stuck on by the bee-keeper (Guy Dhuit).
Bottom: The queen and her court; a worker licking the queen (R. H. Noailles).
V. Free building and wax garlands.
Top: Free building by bees using garlands of wax and stocks (R. H. Noailles).
Bottom: Close-up of wax garland (Guy Dhuit).
VI. Adjustments in construction made by bees in joining up male cells and worker cells, and a queen cell.

Top: How bees attempt to solve a complex geometrical problem: adjustment between worker cells and the larger male cells (R. H. Noailles).

Bottom: Royal cell amidst worker cells (Guy Dhuit).

VII. Collection of nest-building material by ants. Ants amidst twigs used for nest construction (Rapho, photo by Robert Doisneau).

VIII. The finished nest and transport of pupae.

Top: A nest of *Formica rufa* (Guy Dhuit).

Bottom: Ant (*Anomma* sp.) carrying pupa (Rapho, photo by P. J. Corson).

IX. Food exchange and care of the young.

Top: One ant gives food to another (Guy Dhuit).

Bottom: Two ants caring for eggs and young larvae (Guy Dhuit).

X. Ants "milking" an aphid and attacking a beetle.

Top: Formica rufa feeding on the sweet secretion from an aphid (Van den Eeckhoudt).

Bottom: Anomma sp. ant attacking a beetle (P. J. Corson).

XI. Ants blocking the entrance to a nest. An exotic species of ant uses its square-shaped head to seal up completely the entrance to a nest.

Top: An observation window shows the sentinel blocking the entrance (Private collection).

Bottom: An ant approaching a nest entrance which is blocked by the head of another ant (Private collection).

XII. Social contacts in ants and hornets building a nest.

Top: Social contact between two *Myrmica* ants; no doubt the exchange of information (Van den Eeckhoudt).

Bottom: Hornets building a nest (Guy Dhuit).

XIII. Termite mound in Africa (Rapho, photo by R. Merle).

XIV. Close-up of termite galleries. A close-up showing how termite galleries are constructed: part of the wall of a nest (Private collection).

XV. Cuttle-fish camouflaged, and Siamese Fighting Fish in combat.

Top: Cuttle-fish possess the faculty of merging into the background on which they rest (Rapho, photo by Alain Boisnard).

Bottom: Two fish (*Betta splendens*) fighting (Atlas-Photo, photo by Charles Lenars).

XVI. Sticklebacks in normal and fighting attitudes (R. H. Noailles).

Top: Normal attitude.

Bottom: Fighting attitude.

XVII. Migration of grey geese. Grey geese (*Megalornis grus*) migrating; taken at a great height with a telephoto lens as they passed beneath the moon in the early morning (François Marlet).

XVIII. Flights of starlings.

Seagulls taking off from the Lofoten Islands.

Top: Flight of starlings (*Sturnus vulgaris*) (Eric Hosking).

Bottom: Gulls taking off (*Rissa tridactyla*) from the Lofoten Islands (Atlas-Photo, photo by René Gardi).

XIX. American Robin feeding its young and a fight between a hawfinch and a mistle-thrush.

Top: A mother robin (*Erithacus rubecula*) feeding her young (Atlas-Photo, photo by Paul Popper).

Bottom: A hawfinch (*Coccothraustus coccothraustus*) attacking a mistle-thrush (*Turdus viscivorus*).

XX. A secretary bird attacking a snake. A secretary bird (*Sagittarius serpentarius*) stalking a snake, crest raised and black wings spread (Rapho, photo by Paul Popper).

XXI. Thrush bathing and courting display by black-headed gulls.

Top: Song thrush (*Turdus ericetorum*) bathing (Atlas-Photo, photo by François Merlet).

Bottom: Courting display of black-headed gull (*Larus ridibundus*) on newly built nest (François Merlet).

XXII. Sandpiper in winter and breeding plumage.

Top: Sandpiper (*Tringa erythropus*) in winter plumage (François Merlet).

Bottom: with breeding plumage (François Merlet).

XXIII. A terns display to its mate on the nest and the sitting bird leaves the nest.

Top: A pair of *Chlidonias hybrida* (the white-winged black tern) on the nest.

Bottom: One bird relieves the other on the nest.

XXIV. Tern offering a fish to its mate. The male tern (*Sterna albifrons*) offers a fish to his mate while she is sitting on the eggs.

XXV. A penguin rookery and a file of young penguins.

Top: A "penguin island" in the Pacific: inhabited mainly by Snares Island penguins (*Eudyptus robustus*) (Rapho, photo Norman Laird).

Bottom: File of young penguins (Rapho, photo Sven Gillsäter.)

XXVI. A penguin's greeting. A penguin greeting its mate with a stately bow. (Rapho, photo Erik Parbst).

XXVII. Penguins exchanging incubation duties.

Left: A female gentoo penguin (*Pygoscelis papua*) calls her mate to take over incubation of the egg. (Atlas-Photo, photo by Paul Popper).

Right: Male Antarctic penguin (*P. adeliae*) approaches his mate for the purpose of taking over incubation duty (Atlas-Photo, photo by Paul Popper).

XXVIII. A love duet and maternity among penguins.

Left: A love duet in Antarctic penguins (Atlas-Photo, photo by Paul Popper).

Right: A female Antarctic penguin protects her newly hatched chicks with her outstretched wings (Atlas-Photo, photo by Paul Popper).

XXIX. A herd of black antelopes and one of black buffaloes.

Top: Black antelopes galloping, Rhodesia (Atlas-Photo, photo by Paul Popper).

Bottom: Black buffaloes, Zambia (Atlas-Photo, photo by Paul Popper).

XXX. Gambolling goats and cows fighting.

Top: Gambols among goats (Atlas-Photo, photo by Tobias Bjorn).

Bottom: A fight among cows, Switzerland: the cow on the left is impatient to fight and is rubbing her head on the ground. Up to the present her adversary does no more than peacefully crop the grass and does not seem to be willing to fight (Atlas-Photo, photo by Paul Senn).

XXXI. Sociability among antelopes and Indian buffalo with its calf.

Top: Social contacts between different species (Rapho, photo by Kay Lawson).

Bottom: Cow buffalo and calf, Borneo (Atlas-Photo, photo by G. Tomisch).

XXXII. Gibbon calling (Rapho, photos by Yela).

Left: In repose.

Right: Gibbon calling.

BIBLIOGRAPHY

It is difficult to give a full bibliography here as it would take up too much space. It is, on the other hand, easy to direct my readers towards a series of basic works which will, in turn, lead them to the more specialized books.

For general zoology the standard work, unique in the world, is Grassé's *Traité de Zoologie* (Masson). This treatise, which will soon be completed, has some twenty volumes, each of more than a thousand pages. It is an irreplaceable source book, not only on matters of anatomy and growth but also on behaviour of animals.

For mammals in general the reader should consult that excellent work by F. Bourlière, *Mammals of the World* (Harrap, 1955).

For birds consult P. Barruel, *Birds of the World* (Oxford University Press, 1954), or Armstrong's excellent book, which I have much used, *Bird Display and Behaviour* (London: Constable, 1965; New York: Dover Publications).

For batracians see Angel, *Vie et mœurs des Amphibiens* (Payot).

For insects see Chauvin, *Vie et mœurs des Insects* (Payot, 1956).

On animal instinct in general see Tinbergen, *Social Behaviour in Animals* (Methuen, 1965) and the *Colloque international sur l'instinct* (Singer Polignac Foundation 1954, Masson 1956).

Most of the illustrations in this book have been taken from one of the great journals given below; all are very important: *Zeitschrift Für Tierpsychologie*, edited by the two celebrated zoopsychologists Koehler and Lorenz (Parey, Berlin and Hamburg), *Animal Behaviour*, edited by Worden, Weiskrantz and Aronson (Ballière, Tindall and Cox, London), and *Behaviour*, edited by Baerends and his colleagues (Brill, Leyden, Netherlands).

As regards social insects in general, there is a good international review published in France: *Insectes sociaux* (Masson).

For bees consult the *Annales de l'abeille*, published by the Institut National de la Recherce Agronomique, 149, Rue de Grenelle, Paris.

For social behaviour in general see the conference given at the Centre National de la Recherce Scientifique, *Structure et physiologie des sociétés animales.* 1950–52 (C.N.R.S., 13 Quai Anatole France, Paris).

The following are also recommended: H. Hediger, *Psychology of Animals in Zoos and Circuses* (London: Butterworth, 1950, New York: Collier); H. Piéron, *The Sensations* (J. G. Miller, 1952); Lorenz, *Instinctive Behaviour* (Methuen).

INDEX

INDEX

Albrecht, 145, 147

Alexander, on crickets, 176
  on hierarchy, 213
  on insect song, 180–3

Allee, 217

Anderson, 187

Antelopes, relationship with baboons, 253

Ants:
  Anergates, 122
  appetites, 111–12
  Argentine, 95
  army, 115–17
  Atta, see leaf-cutting
  birds' attraction towards, 222
  co-operation, 101–2
  Dendrolasius, 122
  "fire", 95
  fungus-growing, *see* leaf-cutting
  harvesting, 120–1
  hibernation, 114
  indestructability of, 95
  isolation, effect of, 55
  leaf-cutting, 95, 121–2
  "Lomechusamania", 124–5
  Messor, see harvesting
  mutual relationships, *see* polycalism
  nest building and cleaning, 103–11
  Oecophylla, see weaver
  parasol, *see* leaf-cutting
  polycalism, 98–9
  Polyctena, 100
  Polyergus, 123–4
  population, 112
  red, 96–7
  rufa, 100–1
  slave-making, 122–4

Solenopsis sae vissima, see "fire" ant
  Tetramorium, 122
  weaver, 117–19

Apis dorsata, 18

Apis flora, 18

Apis Mellifica, see Bees

Armstrong, 198–9, 203, 205, 207, 219

Baboons:
  de-lousing, 254
  diet, 262
  dominance, 255, 263
  economic dependence, 262
  friendships, 254
  group size, 262
  guard duty, 256–8
  inter-communication, 263
  mother-child relationship, 263
  play, 254, 263
  population structure, 262
  sexual habits, 255–6, 263
  social life, 253, 262
  territory, 262
  tribal life, 252–6

Baerends, 169

Becker, 42

Beebe, 203

Bees:
  acariosis, 50–3
  adaptation, 80–1
  aggressiveness, 66
  Andrena, 17
  building 69–80
  cells, 69–80
    modification of, 78–80
  colonies, 17–18
    cohesion and defence of 63–8
  comb, 21, 69–80

Encyclopaedia of Tropical Fish

REGINALD DUTTA

This book is a complete and authoritative guide to an increasingly popular hobby. Not only does it list and describe a wide variety of tropical fish—half of which have never before been listed elsewhere: It also provides the collector with full information on how to buy correctly and maintain his fish trouble free.

The section on Diseases, for instance, gives advice based on cures of thousands of fish by the author and his staff—the product of thirty years of original research.

Reginald Dutta is the Managing Director of Fish Tanks Ltd. and the author of several books on pet fish.

BOOKSHELF SPECIAL!

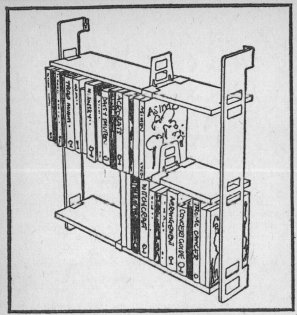

ONLY 132½p (26s 6d) Plus 24p (4s 9d) postage and packing

This three-shelf bookshelf is specially designed to fit your paperbacks. There's room for up to 120 standard size paperbacks. Each shelf has its own adjustable bookend to hold the books snugly in place.

It's handsome! It's made of very strong, very lightweight metal. The shelves are beautifully finished in matt white, yellow and red; sides and bookends in gunmetal grey.

It fits anywhere! Height 21″, width 19″, depth 4⅓″ (approx). You can fix it at floor level or eye level, above a bunk or beside a divan. It's easy to assemble: full instructions are enclosed.

You save almost the price of a fat new paperback! Order here and now on the coupon—and this bookshelf costs you only 132½pp (26s 6d) plus 24p (4s 9d) postage and packing. 156½p (31s 3d) altogether. This represents a saving of 34½p (7s) on the manufacturer's recommended price 191p (30s 3d) for a similar shelf in the shops.

Please fill in both parts of the coupon and send it with cheque or postal order, made payable to Sphere Bookshelf Offer to:
**Sphere Bookshelf Offer,
Orient House, Granby Row,
Manchester M1 7AU**
This offer is open only to readers in Gt. Britain and N. Ireland. Closing date June 30th 1972.

Allow 21 days for delivery.

All Sphere Books are available at your bookshop or
newsagent: or can be ordered from the following address:

Sphere Books, Cash Sales Department,
P.O. Box 11, Falmouth, Cornwall.

Please send cheque or postal order (no currency), and allow
9d. per book to cover the cost of postage and packing
in U.K., 1s. per copy overseas.